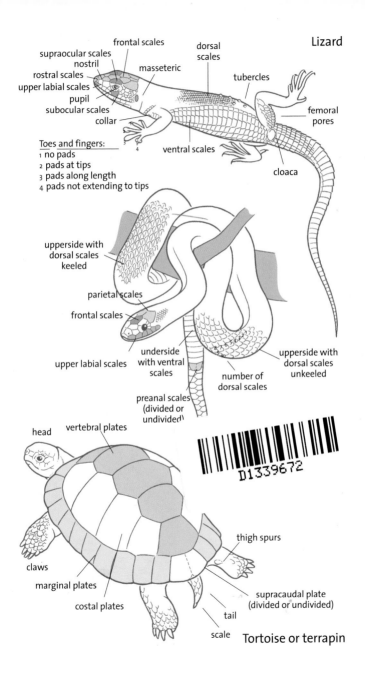

Lizard

frontal scales
dorsal scales
supraocular scales
nostril
masseteric
tubercles
rostral scales
upper labial scales
pupil
subocular scales
collar
femoral pores
ventral scales
cloaca

Toes and fingers:
1 no pads
2 pads at tips
3 pads along length
4 pads not extending to tips

upperside with dorsal scales keeled
parietal scales
frontal scales
upper labial scales
underside with ventral scales
number of dorsal scales
upperside with dorsal scales unkeeled
preanal scales (divided or undivided)

head
vertebral plates
thigh spurs
claws
marginal plates
costal plates
supracaudal plate (divided or undivided)
tail
scale

Tortoise or terrapin

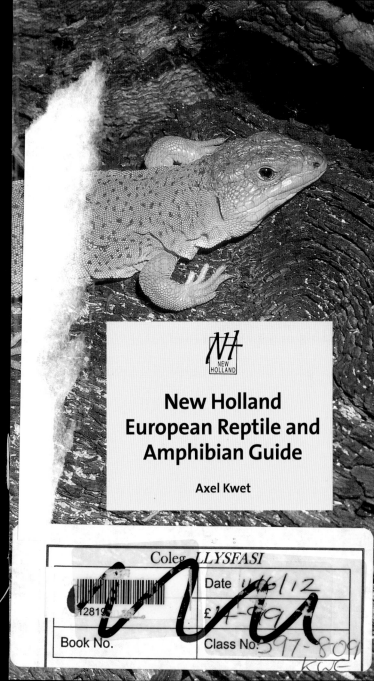

New Holland
European Reptile and
Amphibian Guide

Axel Kwet

First published in 2009 by New Holland Publishers
London ● Cape Town ● Sydney ● Auckland
www.newhollandpublishers.com

Garfield House, 86-88 Edgware Road, London W2 2EA, United Kingdom
80 McKenzie Street, Cape Town, 8001, South Africa
Unit 1, 66 Gibbes Street, Chatswood, NSW 2067, Australia
218 Lake Road, Northcote, Auckland, New Zealand

10 9 8 7 6 5 4 3 2 1

ISBN 978 1 84773 444 0

Although the publishers have made every effort to ensure that information con-
tained in this book was meticulously researched and correct at the time of going
to press, they accept no responsibility for any inaccuracies, loss, injury or incon-
venience sustained by any person using this book as reference.

Publisher: Simon Papps
Publishing Director: Rosemary Wilkinson
Editors: Simon Papps, Beth Lucas and Julia Bruce

Copyright © 2009 New Holland Publishers (UK) Ltd including the year of first
publication. Copyright © 2005: Franckh-Kosmos Verlags-GmbH &Co. KG,
Stuttgart. Original title: Kwet, Reptilien und Amphibian Europas
Printed in the Czech Republic

Images

(a=above; b=below; l=left; r=right). Akeret (pp. 58, 68, 88, 119a, 137, 150, 168, 176, 191,
193b, 210, 211, 221, 228b, 239 both, 243), Amat (pp. 23a, 51b, 67, 147b, 155, 159 both,
235b, 242), Aquilar (p. 63b), Billing (p. 237b), Bischoff (pp. 22a, 152, 153, 177, 188, 189),
Böhme (pp. 56, 57a, 71b, 79a, 131b, 133b, 145a, 151, 157a, 167b, 175, 183b), Carretero (p.
203), Grabert (pp. 107a, 127b, 162b, 172, 229a), Hallermann (p. 185), Hecker/Sauer
(pp. 72, 73 both, 74, 128, 202), C. König (pp. 66, 200), R. König (p. 55), Kühnel (p. 45),
Martinez-Solano (pp. 22b, 26, 29 both, 44, 51a, 64, 147a, 165, 215), Nöllert (pp. 23bl, 25
both, 27, 34, 38, 46, 52, 53a, 57b, 61 both, 62, 65, 69, 79b, 81a, 85b, 89 both, 95b, 101,
102, 105a, 112, 119b, 120, 122, 129, 132, 133a, 134, 135a, 138, 141b, 142, 144, 154, 156, 157b,
160, 161a, 164, 169, 171b, 173, 175a, 178, 179 both, 182, 184, 192, 194, 195, 197b, 199b, 204,
205, 206, 207 both, 213 both, 220, 226, 227 both, 229b, 231b, 234, 236, 238, 240), Pauler
(p. 23br), Rivera (pp. 32, 33, 50, 54, 146, 166, 167a, 190), Rödel (pp. 28, 108, 241a), Solé
(pp. 31a, 149a, 183a, 193a, 214, 222, 223b, 225b, 228a, 233b), Stephan (pp. 92a, 139b),
Vetter (p. 237a), Ziegler (pp. 47b, 235a), all other images by author.

Artwork by Wolfgang Lang.

Photograph on pp. 2-3 is an Ocellated Lizard (*Timon lepidus*)

Photograph on pp. 18-19 is a Pool Frog (*Rana lessonae*).

Contents

Acknowledgements

I would like to thank all my colleagues, friends and relatives who have always supported me during my research on amphibians and reptiles. My special thanks go to Steffi Tommes who encouraged me to begin the project and who completely committed herself to the realization of this book. I would like to dedicate my thanks to my place of work, the Museum of Natural History, Stuttgart, and to its curator of herpetology, Dr. Andreas Schlüter, for the opportunity to write this book. For taking a critical look through the first drafts of this book I owe many thanks to Andreas Nöllert, Bärbel Oftring, Dr. Jörg Plötner, Dr. Andreas Schlüter and Wolfgang Bischoff. Melanie Wiegand-Brauner created the distribution maps, Wolfgang Lang did the line drawings, and Stefan Hecht gave tremendous support to the project. I would also like to thank all photographers who donated photographs to improve and illustrate the book. Among many others I would especially like to name Andreas and Christel Nöllert as well as (in order of number of photographs included) Dr. Mirco Solé, Dr. Beat Akeret, Iñigo Martínez-Solano, Prof. Wolfgang Böhme, Xavier Rivera, Felix Amat, Wolfgang Bischoff, Markus Grabert, Günther Stephan, Dr. Mark-Oliver Rödel, Dr. F. Sauer and Frank Hecker, Ferran Aguilar, Dr. Jakob Hallermann, Prof. Dr. Claus König, Dr. Thomas Ziegler, Dr. Herbert Billing, Miguel Carretero, Klaus-Detlef Kühnel, Ingo Pauler, Harald Aberle, Dr. Salvador Carranza, Dr. Joachim Horstkotte, Dr. Alex Kupfer, Dr. Ulrich Roser, Dr. Andreas Schlüter, Dr. Günther Steinbrück and Dr. Hans-Georg Wolf. Selecting photographs turned out to be extremely difficult and unfortunately it was not possible to use all available photographs, for which I apologize. My special thanks go to my girlfriend Ute Schweizer who has always shown understanding for my hobby, which has become a profession.

Common Newt

About this book

Detailed accounts cover 143 of the most important European species of amphibians and reptiles. Numerous photographs depict all northern and central European and most of the Mediterranean species and subspecies, each of which is dealt with in one to four pages. A further 59 species are covered by concise accounts, and clear distribution maps show the ranges of almost all 202 species.

Common Midwife Toad

Protection of reptiles and amphibians

In recent years global interest in and concern about reptiles and amphibians has increased enormously, not only among biologists but also in the media. The advertising industry uses frogs as cute symbols, snakes represent danger and tortoises are deemed to be wise. Many species have a more direct relevance to humans as the subjects of scientific research or as providers of food and leather. Others are used for pest control or produce poisons and other pharmacologically effective secretions. Amphibians and reptiles are important parts of food chains and ecosystems and at the same time they are vital indicators of the health of our environment.

The protection of these animals is not only sensible for ecological, aesthetic and moral reasons, but it also serves our own economic needs. The worldwide extinction of species has been affecting reptiles and amphibians for a long time. This is not only due to direct changes such as the destruction of habitats, the use of pesticides, increasing traffic or hunting and trapping. All of a sudden certain amphibians, such as the Golden Toad from Costa Rica or the Gastric Brooding Frog from Australia disappeared from habitats completely untouched by humans. Some of these mysterious population collapses may be related to global warming or increasing UV radiation; others can be put down to infestation with fungi (chytridiomycosis), bacteria or iridoviruses. In most cases the exact reasons have yet to be identified, but scientists all over the world are currently making enormous efforts to find out why so many amphibian species are disappearing. Herpetologists form international teams under the supervision of the International Union for the Conservation of Nature (IUCN), the Declining Amphibian Population Task Force (DAPTF) and the Global Amphibian Assessment (GAA).

In Britain the National Reptile and Amphibian Recording Scheme (NARRS; www.narrs. org.uk) runs volunteer-based surveys to monitor the conservation status of all UK species; the Amphibian and Reptile Groups of the UK (ARG-UK; www.arg-uk.org.uk) operates a network of wildlife volunteer groups concerned with the conservation of native species; and the Herpetological Conservation Trust (HCT; www.herpconstrust.org.uk) aims to safeguard Britain's threatened herpetofauna through research and educa-tion, and it also manages several nature reserves in southern England specifically for the rarer species.

Female Slow Worm with young

This book is not just meant to

help nature enthusiasts and biologists with an interest in herpetology to identify species; it also aims to support an increasing general interest in reptiles and amphibians. It may even change the opinion of those people who still agree with the famous Swedish biologist Carolus Linnaeus (1707–1778) for whom amphibians and reptiles were 'disgusting, nasty, pale, dirty, impetuous and calculating' animals 'with a disgusting smell and repulsive voice'.

Geographical scope

Europe forms just a small part of the huge landmass of Eurasia. Usually the border between Europe and Asia is defined as a line that runs from the Urals to the Caucasus, across the Black Sea and along the Turkish coast of the Mediterranean Sea towards the south. Within this area more than 200 species of amphibians and reptiles are found. In a pocket guide intended for use in the field it is not possible to fully describe all species that occur in this area, sometimes in extremely small populations. Species which occur only in areas which have not yet been well developed for tourism, for example the Caucasus and the Crimea, are not treated fully in this book. The eastern border chosen for this book is a line that runs from the North Cape along the 25th degree of longitude over the Carpathian Mountains and Rhodope Mountains to the Aegean Sea and from there along the Turkish coast southward. The reptiles that occur on the bigger Mediterranean islands such as Corsica, Sardinia, Sicily, Malta and Crete are described in this book, as well as those reptiles that can be found on the Balearic Islands and most of the Aegean Islands off the Turkish coast that politically belong to Greece. Not included are those species that only occur in Asia Minor or on the islands that geographically belong to Africa, such as Madeira, the Canary Islands, the Azores and the Salvage Islands.

Species accounts

About 190 species of reptiles and amphibians – including the only two species of turtles breeding in Europe – inhabit the area described above. Of these, the 133 most important species are covered in detailed accounts. The British species and those which are widely distributed across Europe or have a greater number of subspecies are more thoroughly treated than those with just a tiny distribution. Amphibians and reptiles that are not mentioned due to space limitations are dealt with, and in most cases illustrated, under the species accounts of their closest relatives; their distributions are shown together in one map. Coloured bands at the top of the page show the division of reptiles and amphibians into five main groups (which reflect the scientific nomenclature of order and suborder). Line drawings in the coloured bands represent the 25 families occurring in Europe. A key to identification (beginning on p. 12) based on easily recognizable external features enables identification up to genus, and further identification is possible by checking the species accounts.

The scientific names are in line with the latest taxonomic and systematic knowledge (with the exception that we keep the widely used definition of 'reptiles', which is out of date in terms of evolutionary biology). Some traditional names had to give way to the new and unknown, but the older names of genus and species are also given in the text. Especially among snakes of the 'old' genera *Elaphe* and *Coluber* new names have been used. The scientific name of the Aesculapian Snake has changed from *Elaphe longissima* to *Zamenis longissimus*, whereas the scientific name of the Western Whip Snake has switched from *Coluber viridiflavu*

to *Hierophis viridiflavus*. The different species of lizards formerly lumped under the genus *Lacerta* now belong to the genera *Archaeolacerta* and *Iberolacerta*. The names *Podarcis* for the Wall Lizards, *Timon* for the Ocellated Lizards and *Zootoca* for the Viviparous Lizard have already been largely established. The new name for the European Leaf-toed Gecko, formerly known as *Phyllodactylus europaeus*, is *Euleptes europea*, and the name for the European Glass Lizard has changed from *Ophisaurus apodus* to *Pseudopus apodus*. Other changes to the 'old' nomenclature have become necessary because the latest scientific findings – especially in the fields of molecular genetics and bioacoustics – have in some cases led to the conclusion that a species has to be split. One example is the Green Lizard, which is now represented by two species (*Lacerta viridis* and *Lacerta bilineata*) and another is the Aesculapian Snake which has been split into *Zamenis longissimus* occurring in central Europe and *Zamenis lineatus* from southern Italy. Crested newts, parsley frogs, three-toed skinks and four-lined snake represent similar examples. The number of splits has increased steadily in recent years.

The different colours in the

distribution maps make it easy to see where the species occur. The distribution of the main species is marked red, whereas the distributions of additional species named in the text are marked in different colours. Next to the distribution maps three details are given which help to identify the main species:

1. The average to maximum body length of fully grown adults. Measurements for frogs and toads are taken from the tip of the head along the back; for turtles and tortoises from their carapaces; and for lizards, snakes and tailed amphibians along their complete length. In case of obvious differences in the body length between males and females, the relevant measurements are given.

2. The family name of the group to which the species belongs. Usually this is the genus, but sometimes it is a more closely related 'part group' of a genus consisting of many species.

3. Two or three key identification points for the species.

The main species account is divided into sections: 'Description', 'Distribution and habitat', 'Notes' and, if needed, 'Subspecies'. Species that look alike, and are thus difficult to identify, are treated under 'Similar species', and other related species of amphibians and reptiles without their own full account are treated under 'Additional species'. We have tried to give as much useful information as possible in the limited space available in a pocket guide.

A pair of Eastern Green Lizards

Identification key

Two of the three recognised orders of amphibians occur in Europe: the tailed amphibians (order Urodela) and the tailless amphibians (order Anura). The third order is Gymnophiona (caecilians, or legless ampibians) which is distributed in the tropics. The reptiles are represented in Europe by tortoises (order Testudines) lizards and snakes (order Squamata), but not crocodiles (order Crocodylia) or beaked reptiles (order Rhynchocephalia). Within Squamata, three suborders are distinguished here: lizards (Sauria), worm lizards (Amphisbaenians) and snakes (Serpentes). The most important distinguishing features of these different groups are given in the following identification key. Further keys enable identification to the genus level. The diagrams on the endpapers at the beginning of the book illustrate typical features.

Identification key to orders and suborders

1   Skin moist, many glands, no scales (class Amphibia) **2**
1* Skin dry without glands, covered by scales or horny plates
    (class Reptilia) **3**
2   Frog-like body; stocky body without tail; hind legs strikingly longer
    than front legs .... **Tailless amphibians (order Anura)**
2* Shape like a newt or salamander; body long with tail; all four legs
    about the same length .... **Tailed amphibians (order Urodela)**
3   Body covered by a shell with carpaces covered by large horny plates
    .... **Tortoises, terrapins and turtles (order Testudines)**
3* Body not covered by shell, but with small horny plates (Snakes and
    Lizards, order Squamata) **4**
4   Regular ring-like grooves on body, no legs
    .... **Worm lizards (suborder Amphisbaenia)**
4* No regular ring-like grooves on body; with or without legs **5**
5   Always without legs; no moveable eyelids
    .... **Snakes (suborder Serpentes)**
5* Usually with legs; if legless, then eyelids moveable
    .... **Lizards (suborder Sauria)**

Identification key to the genera of the tailed amphibians (order Urodela)

1   Body eel-like; eyes atrophied; just three toes on front legs and two
    toes on hind legs .... **Olms (*Proteus*), p. 55**
1* Body salamander- or newt-like; eyes not atrophied; four toes on front
    legs and four or five toes on hind legs **2**
2   Four toes; underside of tail bright red
    .... **Spectacled Salamander (*Salamandrina*), p. 27**
2* Five toes; underside of tail not bright red **3**

3 toes short, strong and connected by webs; nasolabial groove (visible with lens) .... **Cave salamanders (*Speleomantes*), pp. 52-54**
3* toes normal and not connected by webs, but sometimes bordered by webs; no nasolabial groove **4**
4 Land salamanders; in cross-section, tail oval or round, but not laterally flattened; paratoid glands large, raised and porous.................. **5**
4* In water or on land; tail laterally flattened; paratoid glands small or lacking, not porous **6**
5 Very long tail (1.5 to 2 times body length); short and delicate limbs; usually two golden dorsal stripes .... **Golden-striped Salamander (*Chioglossa*), p. 26**
5* Tail not longer than body; strong limbs; no golden dorsal stripes .... **Typical salamanders (*Salamandra*), pp. 20-26**
6 Large (up to 30 cm) and strong with flattened head and small eyes; laterally a row of prominent warts (the ribs may protrude through these) .... **Sharp-ribbed Newt (*Pleurodeles*), pp. 28-29**
6* Length up to 20 cm, eyes not strikingly small; no row of prominent warts on each flank.................................... **7**
7 Only in or near cold running water in Pyrenees, on Corsica and Sardinia; species in Pyrenees has very rough skin .... **Brook newts (*Euproctus*), pp. 30-33**
7* Usually not in cold, running water, but in a stretch of standing water; not on Corsica and Sardinia. If in Pyrenees, then skin not rough .... **Pond newts (*Triturus*), pp. 34-51**

Identification key to the genera of tailless amphibians (order Anura)

1 Underside with bright yellow, orange or red markings .... **Fire-bellied toads (*Bombina*), pp. 56-61**
1* Underside without bright yellow, orange or red markings **2**
2 Round pads at tips of fingers; back mostly bright green .... **Tree frogs (*Hyla*), pp. 56-61**
2* No round pads at tips of fingers; back coloured in different shades of brown, grey or green... **3**
3 Pupil shape seen in light vertically oval to almost round, but not vertically slit-shaped... **4**
3* Pupil shape vertical when seen in light, either vertically oval, vertically slit-shaped, round triangular or heart-shaped (tip pointing downwards)... **6**
4 Skin dry with lots of warts; stout body with short hind legs; large, obvious paratoid glands .... **Typical toads (*Bufo*), pp. 76-83**
4* Skin moist and smooth; slender body with relatively long, strong hindlegs; no paratoid glands .... **Typical frogs (*Rana*) 5**
5* Dark brown patch on temples in which the ear drums are positioned; eyes well separated; back not green; males with or without inner vocal sacs .... **Brown frogs (*Rana*), pp. 90-101**

5* Usually without patch on temple; eyes close together; back mostly green, sometimes green to brownish, often with bright central line; males with external vocal sacs
.... **Water frogs (*Rana*), pp. 102-113**

6 Black or brown, sharp-edged, prominent spade on base of first (smallest) toe (spade as long or longer than first toe); eardrum not visible .... **Spadefoots (*Pelobates*), pp. 70-74**

6* Small, not sharp-edged spade on base of first toe (spade shorter than first toe); eardrums often visible **7**

7 Pupils seen in light vertical and slit-shaped **8**

7* Pupils seen in light oval, round triangular or heart-shaped (tip pointing downwards) .... **Painted frogs (*Discoglossus*), pp. 67-69**

8 Delicate body with long hind legs; usually with webbing between toes; back often with greenish patches. **Parsley frogs (*Pelodytes*), p. 75**

8* Plump body; no membrane along toes (occasionally hints of semipalmations) .... **Midwife toads (*Alytes*), pp. 62-66**

Identification key to tortoises, terrapins and sea turtles (order Testudines)

1 Front limbs paddle-like, bigger than rear limbs; in the sea or by the coast **2**

1* Front limbs not paddle-like, about the same size as rear limbs; on land or in freshwater **3**

2 Carapace with five (or more) pairs of costal plates (one pair often small) **Kemp's Ridley (*Lepidochelys*) and Loggerhead Turtles (*Caretta*), p. 128**

2* Carapace with four pairs of costal plates. **Green Turtle (*Chelonia*), p. 128**

3 Carapace domed; limbs with claws, but without moveable toes; found on land .... **Tortoises (*Testudo*), pp. 114-119**

3* Carapace low; limbs with claws which are connected to toes by webbing; in or near freshwater (rarely in brackish water) **4**

4 Neck with yellow spots, but without stripes; carapace and undershell not rigid, but connected by a flexible cartilaginous 'bridge' .... **European Pond Terrapin (*Emys*), pp. 124-127**

4* Neck with striking yellow stripes; carapace and undershell rigid and connected by bony 'bridge' **5**

5 Neck with obvious yellow stripes and a clear red patch on cheeks .... **Red-eared Terrapin (*Trachemys*), p. 127**

5* Neck with yellowish stripes and without red cheek patch .... **Old World terrapins (*Mauremys*), pp. 120-123**

Identification key to European lizards (suborder Sauria) and worm lizards (suborder Amphisbaenia)

1 Body segmented by ring-like grooves; snake-like, without legs .... **Worm lizards (*Blanus*), p. 200**

1* Body not segmented by ring-like grooves; with or without legs...... **2**

2   Body dorsally flattened; fingers and toes fused together, like pincers; protrudings eyes, which can be moved independently
.... **Chameleons (*Chamaeleo*), pp. 130-131**

2* Body not dorsally flattened; fingers and toes not fused together; eyes not protruding and cannot be moved independently **3**

3   Pupil vertical, slit-like if seen in light **4**

3* Pupil round if seen in light, not vertical or slit-like **7**

4   Toes with flat, adhesive pads.. **5**

4* Toes without pads and strongly kinked
.... **Kotschy's Gecko (*Cyrtopodion*), p. 136**

5   Toes with undivided, flat adhesive pads along their whole length
.... **Moorish Gecko (*Tarentola*), pp. 132-133**

5* Toes only partly covered by pads.................................... **6**

6   Pads are heart-shaped and only at tips of toes
.... **European Leaf-toed Gecko (*Euleptes*), p. 137**

6* Pads in 2 rows, not reaching tip of toe; tips of toes free and with small claw...................... **Turkish Gecko (*Hemidactylus*), p. 134-135**

7   Upperside of head with irregular small scales, which are similar to back scales; tail spiny .... **Starred Agama (*Laudakia*), p. 129**

7* Upperside of head covered by large scales which are clearly different from back scales.. **8**

8   Body without legs **9**

8* Body with legs (sometimes much reduced) **11**

9   On each side one deep groove that runs from the head to the base of the tail .... **European Glass Lizard (*Pseudopus*), pp. 198-199**

9* No groove on each side of the body **10**

10 Small, brown spots form lines on upperside (each scale with one spot); head almost arrow-like if seen from above; not more than 20 rows of dorsal scales around mid body; only in southern Greece (Peloponnese) .... **Limbless Skink (*Ophiomorus*), p. 195**

10* No lines with small dark spots on upperside; head rounded if seen from above; at least 23 rows of dorsal scales around mid body; in almost all of Europe.................... **Slow worms (*Anguis*), p. 196-197**

11 Body shape like typical lizard; usually strongly developed legs; underside of the thighs with one row of femoral pores **13**

11* Body shape like snake; often with underdeveloped legs; if legs well developed, then without femoral pores **12**

12 Eyelids fused together to rigid capsula (snake eye); very small and delicate (not longer than 12 cm). **Snake-eyed Skink (*Ablepharus*), p. 194**

12* Eyelids moveable; strongly built (often longer than 15 cm)
    Chalcides skinks (*Chalcides*) and Levant Skink (*Mabuya*), pp. 190-193

13 Quite large dorsal scales, strongly keeled and positioned like roof tiles.. **14**

13* Dorsal scales small and only inconspicuously keeled, especially at rear and more or less sitting side by side................................ **16**

14 Eyelids fused together to form transparent 'glasses' (eye does not close if touched carefully) ...... **Snake-eyed Lizard (*Ophisops*), p. 180**

14* Eyelids moveable (eye closes if touched carefully) **15**
15 With well developed, rough collar (a row of big scales, very different
    from much smaller neck scales close to head)
    .... **Algyroides lizards (*Algyroides*), p. 186**
15* Collar absent or barely recognizable
    .... **Psammodromus lizards (*Psammodromus*), p. 181**
16 Subocular scale does not reach upper lip or just reaches it; two
    supraocular scales, no occipital scale
    .... **Spiny-footed Lizard (*Acanthodactylus*), p. 184**
16* Subocular scale reaches upper lip; four supraocular scales; usually
    with occipital scale and collar
    .... **Meadow Lizard (*Darevskia*), rock lizards
    (*Archaeolacerta* and *Iberolacerta*), green lizards (*Lacerta*), wall lizards
    (*Podarcis*), Ocellated Lizard (*Timon*), Common Lizard (*Zootoca*), p. 138-179**
Note: reference to the species accounts is necessary in order to identify
the species of lacertids in these genera.

Key to the Identification of the European groups of snakes (suborder Serpentes)

1   Ventral scales not enlarged; worm-like.. **Worm Snake (*Typhlops*), p. 201**
1* Ventral scales enlarged, forming row of scales **2**
2   Large ventral scales only up to about a third of body width; eyes very
    small and almost on upperside of head; tail very short
    .... **Sand Boa (*Eryx*), p. 202**
2* Broad ventral scales more than half body width; eyes not strikingly
    small and most often positioned on sides of head; tail quite long **3**
3   Dorsal scales strikingly keeled **4**
3* Dorsal scales not keeled (or sometimes weakly keeled).............. **6**
4   Pupil round; pre-anal scale divided; upper jaw without erectile fang
    .... **Grass, Dice and Viperine Snakes (*Natrix*), pp. 226-233**
4* Pupil slit-shaped; pre-anal scale undivided; upper jaw with
    erectile fang **5**
5   Upperside of head with many keeled scales (sometimes bigger scales
    above the eyes)..**Mountain vipers (*Montivipera, Macrovipera*), p. 243**
5* Upperside of head with small, keeled scales and some bigger scales,
    or, if all small scales are keeled, with turned-up snout or nose horn
    .... **Vipers (*Vipera*), pp. 234-242**
6   Pupil vertical and slit-like or vertically oval if seen in light **7**
6* Pupil round if seen in light **8**
7   Head-shape distinct from body; pupil slit-like
    .... **Cat Snake (*Telescopus*), p. 210**
7* Head is not distinct from body; pupil vertically oval; black hood-like
    pattern on nape......... **False Smooth Snake (*Macroprotodon*), p. 215**
8   Obvious protruding supraocular scales form 'strict' expression; long
    and narrow front scale ..... **Montpellier Snake (*Malpolon*), p. 224-225**
8* No obvious protruding supraocular scales; front scale not long
    and narrow **9**

9 Small (not longer than 50 cm) and delicate (as thin as a pencil); uniformly coloured body with dark head-markings; only 15–17 dorsal scales around mid body; occurs on just a few Greek islands
.... **Dwarf Snakes (*Eirenis*), p. 211**

9* Larger and broader; body not uniformly coloured and with dark head-markings; 19–29 dorsal scales around mid body **10**

10 Head with broad, dark, more or less conspicious U-shaped marking on nape and with an obvious stripe running from the eye down to the neck .... **Smooth snakes (*Coronella*), p. 212-214**

10* Without U-shaped marking on nape and no dark line running from eye down to the neck **11**

11 Upperside uniform brown except for a row of dark dots mostly surrounded by a fine white line on neck; slender and graceful
.... **Dahl's Whip Snake (*Platyceps*), p. 204**

11* Upperside not uniform brown with dark dots on side of neck **12**

12 Head often conspiciously distinct from body; long and small tail which has fine dark lines, whereas the rest of the body is dark spotted; 19 dorsal scales around mid-body
.... **Whip snakes (*Dolichophis, Hierophis*), p. 205**

12* Tail without fine streaking (a few broad stripes may occur); 21–29 dorsal scales around mid-body **13**

13 Dorsal scales weakly keeled; four conspicuous dark lines on back or strong markings; only in Italy and the Balkans
.... **Four-lined Snake and Blotched Snake (*Elaphe*), pp. 216-217**

13* Dorsal scales smooth **14**

14 Back with obvious dark pattern resembling rope ladder (juveniles) or two obvious lines (adults); mostly 27 dorsal scales around mid-body; only in Iberia and southern France
.... **Ladder Snake (*Rhinechis*), p. 222-223**

14* Usually without dark back markings formed like lines; 21–29 dorsal scales around mid-body; if there is a hint of dark lines, then just 21–23 dorsal scales around mid-body **15**

15 Head quite obviously distinct from body; long, slender tail; 25–29 dorsal scales around mid body; pattern on back either consisting of dark fringed patches and horseshoe-shaped dark marking on back of head (Horseshoe Whip Snake only in Iberia and Sardinia) or a row of dark blotches which are far apart (Algerian Whip Snake only in Malta; Coin-marked Snake only in Turkey and some Greek islands)
.... **Whip snakes (*Hemorrhois*), p. 203**

15* Head indistinct from strong body; either just 21–23 scales around mid body (aesculapian snakes) or, if there are 25–27 scales around mid body, with striking back pattern consisting of red dots fringed by black lines (Leopard Snake)
.... **Aesculapian snakes and Leopard Snake (*Zamenis*), pp. 218-221**

Note: as the true snakes are not easy to identify with the means of a key alone, comparing the different species accounts is always recommended.

Species accounts of European amphibians and reptiles

Fire Salamander
Salamandra salamandra
Salamanders and newts SALAMANDRIDAE

> 14–25 cm
> True salamanders
> Shining yellow and black upperside; robust build; large paratoid glands with pores

Description Strong and stout land salamander with short limbs, tail round in cross section. Maximum length 25 cm, but in north of range mostly reaching 14–18 cm. Skin quite smooth, two rows of pores along back and one obvious pair of paratoid glands at the rear of the head. Coloration and pattern unmistakable, but very variable depending on sub-species; usually shining black with a pattern of yellow, orange-yellow or reddish irregularly formed patches or stripes. Belly black to grey, often weakly spotted yellow.

Distribution and habitat Widely distributed in western, central and southern parts of mainland Europe. Absent from Britain, Scandinavia and almost all Mediterranean islands. Most commonly found in cool and humid deciduous forests, often close to streams, but also in more open landscapes like meadows or even in towns. Quite common in mountainous landscapes between 600 and 1,000 m. In the Balkans and central Spain occurs up to 2,300 m. Larvae develop in stagnant parts of streams without any fish and also in ponds and mountain lakes.

Notes Terrestrial species mostly active at night, spending the day hidden under stones and dead wood. Diet: slugs, spiders, insects, centipedes and earthworms. Secretes a noxious substance from the skin as a defence against predators. Can reach 10 to 15 years old or up to 50 years in captivity. Mating takes place on land (usually from March to September). The male gets beneath the female, wraps his front limbs around hers and eventually deposits a spermatophore on the ground. The female lowers her cloacal region down onto it and fertilization is internal. Usually

Larva with external gills

early in spring (and rarely in autumn) the female gives birth to 10 to 80 aquatic larvae. Some subspecies, especially in northern Spain, give birth to fully formed young.

Subspecies Until recently, the Fire Salamander was regarded a very variable species with just a few subspecies. Today four species are recognized, two of which occur in Europe (see Corsican Fire Salamander, p. 23). The nominate subspecies *S. s. salamandra* is split into at least 11 subspecies, most of which can be found in Iberia. *S. s. bernardezi*

Fire Salamander of the central Greek subspecies werneri

Fire Salamander of the Iberian subspecies S. s. crespoi

and *S. s. fastuosa* from north-west Spain are very similar, having graceful bodies with prominent yellow stripes. *S. s. almanzoris*, also very graceful, only occurs in the Sierra de Gredos and Sierra del Guadarrama. It shows hardly any yellow on the back at all, apart from a few tiny patches. The closely related *S. s. bejarae*, which lives in eastern and central Spain, is similar; some populations tend to show reddish coloration. *S. s. morenica* from Andalucia and *S. s. gallaica* from Galicia and

central and northern Portugal also have a lot of red in the yellow-black coloration, especially on the head. Closely related to these two subspecies is *S. s. crespoi,* a fairly large subspecies from southern Portugal with a long, flat head and long limbs. *S. s. longirostris* is also large and occurs in the Sierra de Ronda in southern Spain; four prominent yellow patches on the upperside of the pointed head are typical for this subspecies. *S. s. gigliolii* occurs in Italy, south of the plains of the Po Valley. It is quite small, has delicate limbs and is predominantly yellow with black blotches. The distribution of the striped *S. s. terrestris* reaches from northern Spain across France to the west of Germany. By far the largest distribution is shown by the irregularly patterned *S. s. salamandra*, from southern Switzerland, northern Italy, south-eastern Germany,

S. s. bernardezi *from Spain*

Czech Republic, Slovakia and most of the Balkans. There are two forms in the southern Balkans; *S. s. beschkovi* in southern Bulgaria and *S. s. werneri* in central Greece. These forms may represent valid subspecies.

Additional Species

■ **Corsican Fire Salamander**
Salamandra corsica (distribution in blue) was long regarded as a subspecies of the Fire Salamander. It is rather plump and has many small, irregularly arranged orange-yellow spots.

Iberian subspecies S. s. fastuosa

■ **Luschan's Salamander**
Lyciasalamandra helverseni (distribution in green) occurs only on the Aegean Islands of Kassos, Karpathos and Saria. Until recently it was regarded a subspecies of the Lycian Salamander (*L. luschani*) and placed in the genus *Mertensiella*. It is 12–14 cm long, slender, agile and terrestrial with dark brown and yellow spotted uppersides. Males show a unique soft spur above the base of the tail.

Corsican Fire Salamander

S. s. gigliolii *from Italy*

Alpine Salamander
Salamandra atra
Salamanders and newts SALAMANDRIDAE

> 12–15 cm
> True salamanders
> Shining black back; ribs prominent on flanks, obvious rows of warts; large paratoid glands

Description Relatively slender, mid-sized land salamander. Unmistakable because of the uniform, black, shining coloration (except for the rare, yellowish spotted supspecies *S. a. aurorae*). Skin relatively smooth, but with a row of obvious warts along the centre of the back and on the sides of the body. The body appears segmented due to 11–13 costal grooves. Protruding, kidney-shaped paratoid glands. Pointed tail, square in cross section.

Distribution and habitat The Alps and mountains of north-western Balkans. Isolated populations exist south to northern Albania. Found between 430 m and 2,800 m, most common between 800 m and 2,000 m. Prefers moist areas like deciduous forests (especially clearings and forest edges) and wet alpine meadows. Often found under stones and dead wood; locally common.
Notes Slow moving, terrestrial amphibian. Depending on

altitude active between April and October; mostly nocturnal, but it also appears during daylight after rainfall. Diet: insects, larvae, spiders, slugs, earthworms and centipedes. Mating occurs on land (similar to Fire Salamander, see p. 21). Larvae develop in female's body for 2–3 years after which two 4–5 cm long young are born.

Subspecies Aurora's Alpine Salamander (*S. a. aurorae*) occurs exclusively in north-east Italy (Vicenza province between 1,300 m and 1,500 m). It differs from the nominate form by a pattern of variously sized connected yellow spots.

Similar species Crested newts and Alpine Newt can also be very dark on the upperparts but have orange or yellow bellies.

Additional species

■ **Lanza's Salamander**
Salamandra lanzai (distribution

Lanza's Salamander

in blue), is also black but on average a bit larger. It only lives in the Cottian Alps in Western Piedmont (Italy), at altitudes between 1,500 m and 2,000 m. Its tail is longer than that of *S. atra* and rounded at the tip, and there are tiny semipalmations between the toes. The centre of the back is smooth with no pores, the warts on the flanks are long rather than rounded as in the Alpine Salamander.

Aurora's Alpine Salamander

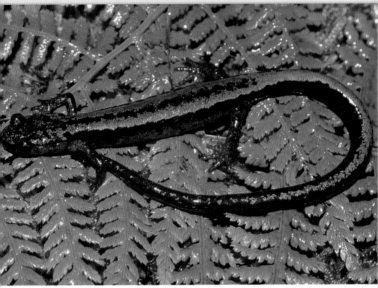

Golden-striped Salamander
Chioglossa lusitanica
Salamanders and newts SALAMANDRIDAE

> 12–16 cm
> Golden-striped Salamander
> Two golden stripes on back; slender body with short legs and long tail

Description Graceful, short-legged salamander with protruding eyes and a long tail making up two thirds of its overall length. Upperside smooth with 10–11 costal grooves, dark brown to blackish and usually with two parallel golden stripes on back, which merge at the base of the tail. Stripes sometimes broken into several spots. Underside dark grey.

Distribution and habitat Northwest of Iberia up to an altitude of 1,300 m. Most common in moist deciduous forests with abundant bushes and mosses, often under stones near rocky streams, or in caves .

Notes Agile terrestrial amphibian. Nocturnal. Capable of swimming if in danger and may shed its tail like a lizard if threatened (autotomy). Can live 8–10 years. Uses sticky tongue to catch invertebrates. Mates on land. Clutch of 10–35 eggs laid between stones in running water or on the walls of caves. Larvae develop in water.

Spectacled Salamander
Salamandrina terdigitata
Salamanders and newts SALAMANDRIDAE

> ♂ 7–9 cm, ♀ 9–11 cm
> Spectacled Salamander
> Pale bar between the eyes; underside of tail and legs red; only four toes

Description Graceful land salamander with a very slender profile. Conspicuous costal grooves, protruding eyes and long tail. Skin quite rough, brown to blackish. Head with yellowish, sometimes diffuse bars between eyes ('spectacles'). Belly whitish with dark patches, throat black. Cloacal region, underside of limbs and tail are bright red. Only four toes on hind legs.

Distribution and habitat Patchily distributed in the western parts of the Apennines (Italy), in humid valleys and shady, overgrown hillsides at altitudes between 200 m and 1,200 m. Most often found near streams, in dense vegetation, under leaf litter, dead wood or stones.

Notes Nocturnal and terrestrial. Often stays hidden. Curls up its tail and displays red underside when in danger. Mating takes place on land in spring. Clutches containing about 30–50 eggs are laid in water; larvae take 2–3 months to develop.

Sharp-ribbed Newt
Pleurodeles waltl
Salamanders and newts SALAMANDRIDAE

> 20–30 cm
> Ribbed newts
> Skin warty; broad, flat head; eyes pointing upwards;
  ribs can clearly be seen on sides

Description Largest and strongest newt in Europe. Flat, broad head with small eyes pointing upwards, no paratoid glands. Skin very rough, ribs can clearly be seen bulging through the flanks; bulges are brown or yellowish. Ribs often project right through the skin, protecting the animal against predators. Upperside grey, brownish, yellowish or olive, mostly with indistinct dark patches. Belly yellowish or light grey, irregularly spotted. Under-side of the dorsally flattened tail is yellowish and shows a low crest. Males have dark rough nuptial pads on the insides of the forelegs during breeding season.

Distribution and habitat Central and southern Iberia. Particularly favours humid, densely vegetated areas up to an altitude of 1,500 m, but also occurs in drier and more open cultivated land. Ponds, wells or slow flowing streams are favoured for breeding; sometimes

28

lives in water all year round. On land it hides under stones or dead wood, or in crevices. **Notes** Mostly nocturnal, but during mating in spring may be active during the day. Diet: insects, slugs and worms. Mating takes place in water. The male grips the arms of the female from beneath for hours or even days, before finally depositing a spermatophore. When releasing her he guides her with a circular movement across the deposited spermatophore, so that she can take it into her cloaca. Fertilization takes place internally. Over a period of weeks 250 to 500 eggs are scattered on aquatic plants in groups of

The ribs project through the skin in some individuals

about 20. The larvae hatch after 8–10 days and metamorphose within a period of about three months, depending on the temperature of the water. Sharp-ribbed Newts are usually active all year round but may slow their metabolism when conditions are very dry.

Sharp-ribbed Newts are predominantly aquatic

Pyrenean Brook Newt
Euproctus asper
Salamanders and newts SALAMANDRIDAE

> 10–16.5 cm
> Brook newts
> Toes with black, horny tips; skin warty; found only in
   the Pyrenees

Description Relatively large brook newt with broad, flattened head, which appears square due to blunt snout. Head is separated from neck by an obvious wrinkle. No paratoid glands. Skin very rough with many tubercles. Tail compressed from side and without tail-crest. Toes with black, horny tips. Upperparts mostly dull, light grey to dark grey, brown-grey, olive-brown or almost black. Immatures and adult males often have yellowish markings, which may merge to a pale central dorsal line. Throat orange with dark brown spots. Belly smooth, strongly coloured orange-red in males and paler and more yellowish in females. Cloacal swelling rounded in males and conical in females, pointing backwards.
Distribution and habitat Only in the French and Spanish Pyrenees and adjacent mountains. Most common at altitudes between 1,000 m and 2,000 m (not higher than 2,550 m) but

may occasionally occur at altitudes of 250 m. In spring and summer found between stones in quiet parts of fast-flowing streams with little vegetation, in ditches and in puddles on the banks of mountain lakes. Requires constant water temperatures below 15°C, so at lower altitudes restricted to streams in caves. Hibernates on land close to streams, often forming large groups.

Notes Diet consists of aquatic insects and larvae as well as small worms and crustaceans. Trout is a significant predator. Internal fertilization takes place in the water. Mating in spring and autumn begins with a display posture by the male in which the tail is raised, showing the orange underside. During mating, which can last for several hours, the male embraces the female with his

Individual with clearly marked line on back

tail and front legs. He releases several spermatophores and massages them into the female's cloaca with his hind legs. Between 30 and 70 eggs are laid and attached to rock surfaces in the stream. After 4–5 weeks the larvae hatch and take about one year to metamorphose. The animals reach maturity after 3–4 years. Pyrenean Brook Newts can can live for up to 20 years.

Pair of Pyrenean Brook Newts in amplexus

Corsican Brook Newt
Euproctus montanus
Salamanders and newts SALAMANDRIDAE

> 7–11 cm
> Brook newts
> Small but obvious paratoid glands; throat not spotted;
> skin rough; only in mountain streams on Corsica

Description Smallest European brook newt, with rather rough skin, a flattened head and small, obvious paratoid glands. Upperside brownish, greenish or grey, mostly dark patterned and often with a reddish, partly interrupted line on back. Belly brownish to light grey, unspotted. Male has spurs on hind legs.

Distribution and habitat Only on Corsica, especially in mountains up to 1,500 m, but not higher than 2,100 m. Rarely found in cool freshwater of lower altitudes. During breeding season found in ditches or slow-flowing streams and the banks of smaller lakes and ponds. Hides under stones when on land.

Notes Active night and day during spring breeding season, living under water. Hibernates on land. Can be found on land during the summer. Mating as Pyrenean Brook Newt (see p. 30). Female attaches eggs to the surfaces of underwater stones.

Similar species Corsica Fire Salamander is much brighter.

Sardinian Brook Newt
Euproctus platycephalus
Salamanders and newts SALAMANDRIDAE

> 10–14 cm
> Brook newts
> Indistinct paratoid glands; pike-like head; throat spotted;
>   skin smooth; only in mountain streams on Sardinia

Description Middle-sized brook newt with flattened, pike-like head, smooth skin, and protruding upper jaw. Indistinct paratoid glands. Upperside reddish-brown to greenish, mostly dark spotted, reddish vertebral line. Belly whitish to brownish. Dark spots on throat and centre of belly especially in males, which also have hind-leg spurs.

Distribution and habitat Only on Sardinia, especially mountains up to 1,800 m and coastal plain. Spring and summer in streams, ponds, lakes and ditches. Found under stones close to the water.

Notes Lungs reduced in size, tolerates higher water temperatures than other brook newts. Mates in water. The male holds the female in his mouth, placing his spermatophores into her cloaca. Between 50 and 200 eggs are attached to the surface of stones over a period of several weeks.

Similar species Gené's Cave Salamander has webbed toes (see p. 54) and is terrestrial.

Alpine Newt
Triturus alpestris
Salamanders and newts SALAMANDRIDAE

> ♂ 8–9 cm, ♀ 10–12 cm
> Pond newts
> ♂ upperside blue, ♀ upperside brownish; belly orange, unspotted; head has no stripes or furrows

Description Mid-sized, flat-headed pond newt with short legs and dorsally flattened tail. Surface of skin smooth while living under water but rather rough and velvety while on land. During courtship males have a narrow, even body crest that merges with the upper tail crest. Upperside of male is blue to greyish blue, bordered by a silvery white band with many black spots on sides of head and flanks. Underneath this is a bright blue band that borders

the orange belly. Back crest also pale with small black spots. Upperside of females brownish, grey, greenish or almost black, often with obvious dark bars. Belly of males and females is unspotted and uniform orange to orange-yellow. Some populations have dark markings on the throat.

Description and habitat Widely distributed, especially in hilly regions of central Europe, from western Ukraine and Romania to northern Spain, and from southern Denmark to Greece and central Italy (with isolated populations in southern Italy). Most common at altitudes between 500 m and 2,000 m, in mountains up to 2,500 m, only rarely found in lowlands. Prefers moist, cool areas close to ponds, for example deciduous forests or well-vegetated valleys, but also gardens and cultivated areas. During breeding season found in ponds or very small patches of standing water such as ditches or large puddles. Can also be found in mountain lakes or slow-flowing streams. In the southern parts of its range sparsely vegetated mountain lakes and barren mountainous areas are favoured. Hibernates on land in crevices where it can gather in large groups. Rarely spends the winter under water.

Neoteny in Alpine Newt of Greek subspecies T. a. veluchiensis

Notes Nocturnal newt, only active during the day in breeding season from late February to June. Display behaviour is quite prolonged. The male gets close to the female and directs pheromones from his cloaca to hers with his tail turned sideways. The male

Female of nominate subspecies

Male of the nominate race

patiently follows the female until she touches a part of his tail to signal her willingness to mate. After the male has deposited a spermatophore the female creeps over it and picks it up in her cloaca (fertilization is internal). Over a period of several weeks the female deposits up to 250 eggs onto aquatic plants, using her hind legs to wrap them in the leaves. The development of the embryos depends on the water temperature and usually lasts 2–3 weeks. The larvae take another three months to mature. Adults become terrestrial in June or July, whereas the young newts leave the water in September and October. Sometimes neoteny occurs: larvae do not metamorphose into adults and gills are still present. This is especially common in Balkan populations in high altitude mountain lakes.

Subspecies The nominate subspecies *T. a. alpestris* is most widespread. *T. a. cyreni* is found in northern Spain. The especially colourful *T. a. apuanus* is found in northern Italy. *T. a.*

Example of terrestrial coloration

inexpectatus has an isolated population in southern Italy and *T. a. lacusnigri* is in the Julian Alps. Some subspecies from the Balkans have been described, but their status remains unclear. *T. a. reiseri* from Bosnia is no longer regarded as a valid subspecies, but *T. a. montenegrinus*, *T. a. piperianus* and *T. a. serdarus* from different mountain regions of Montenegro and *T. a. veluchiensis* from northern and central Greece are still considered valid.

Similar species Terrestrial individuals might superficially resemble Alpine Salamander or crested newts but do not

Male showing aquatic coloration

show a uniform underside. Montandon's Newt shows a uniform orange underside but has obvious furrows on head.

Underside of Alpine Newt of the Greek subspecies T. a. veluchiensis

Common Newt
Triturus vulgaris
Salamanders and newts SALAMANDRIDAE

> 6–11 cm
> Pond newts
> Throat and belly orange, dark spotted; ♂ with tall crest which is undivided between back and tail

Description Slender and delicate pond newt with longish head and smooth skin. Individuals from southern Europe are about 6–9 cm long, smaller than those living in northern and central Europe. Upperside of head has three grooves and five dark stripes (the intensity of which varies between individuals) that are divided by pale lines. Upperside pale brown, yellowish, dark grey or greenish brown; males often with big, dark spots and roundish patches, whereas females show smaller spots or are almost plain. In males, centre of belly bright orange or yellow with brown patches, slightly larger on the sides. Belly of the female is paler and with smaller dark spots. Throat usually spotted dark, but some individuals, especially females, do not show any patches at all. Males from northern and central Europe show an undulated or notched body crest during the spring breeding season that merges into the tail crest without an obvious gap. This crest becomes smooth-edged during the period when the males change from aquatic to terrestrial life. The southern European populations possess a smooth-edged crest year-round. The underside of the tail is coloured a striking orange and blue. Females lack a body crest, and have only a modest tail crest. Tail is quite pointed but only the southern European populations show a tail filament. Males have clearly webbed dark brown toes; males also have a widened cloacal region.

Common Newt (T. v. vulgaris) showing terrestrial coloration

Distribution and habitat An adaptable species widely distributed throughout the whole of Europe apart from the Iberia and some parts of southern France and northern Scandinavia. It is a mainly

Spotted underside of male Common Newt (T. v. vulgaris)

Male of nominate subspecies (T. v. vulgaris) in aquatic coloration

lowland species, occurring in altitudes 100–1,000 m, in some parts of Austria up to 2,000 m. Breeds in small, still, shallow, sunny and weedy ponds with well-vegetated banks, and also ditches, large puddles and small lakes. On land it shelters in moist places, such as under stones or dead wood, in deciduous forests or around ponds. Also found on farmland and in gardens.

Notes Usually less aquatic than other pond newts. Closely related to the Palmate Newt, with which it sometimes inter-breeds. Nocturnal on land, but diurnal in the aquatic breeding season which lasts from March to July. Internal fertilization takes place (see Alpine Newt, p. 36). The females deposit 100 to 300 eggs on the leaves of aquatic plants during one breeding season. The larvae develop quickly in about 6–8 weeks. The young leave the water between July and October and they hibernate along with other newts in crevices or caves. In Britain the species is also known as the Smooth Newt.

Male Common Newt from Greece

40

Subspecies Many subspecies, some of uncertain status, have been described. The large nominate *T. v. vulgaris*, in which the male has an undulating crest during breeding, inhabits most of Europe including Britain. *T. v. meridionalis* occurs in southern Switzerland, Italy and the north-west Balkans. It has an almost smooth-edged crest, as does *T. v. graecus* from the southern Balkans and Greece. Females of the latter subspecies are often finely spotted, while the males show a long tail-filament and deep black webs on the hind toes. *T. v. ampelensis* from Romania, is no longer considered a valid subspecies. Other subspecies occur outside the range covered by this book.
Similar species Palmate Newt lacks dark patches on the throat and belly and the underside of the hind toes show two spots. Italian and Bosca's Newts have

Pair of Common Newts from Greece (male above)

only one facial groove and lack webbing on the toes. Males of these two species, as well as those of Montandon's Newt, lack a body crest and only have a slight tail crest.

Male T. v. meridionalis *from Italy, Switzerland and the Balkans*

Palmate Newt
Triturus helveticus
Salamanders and newts SALAMANDRIDAE

> ♂ 7–8.5 cm, ♀ 8–9.5 cm
> Pond newts
> Throat without spots; ♂ with tail filament, dark webs on toes; two pale patches on underside of hind toes

Description Slender, gracile pond newt. Longish head with three obvious grooves and a dark line that lies between eyes and temples. Upperside pale brown, yellowish or olive with dark patches or weakly barred. Centre of belly silvery orange to yellow, unspotted or with a few indistinct patches. Throat whitish to translucent pinkish and unspotted. During courtship, males have a low, smooth-edged body crest, slight tail crest and obvious tail filament which is about 5–8 mm long and strikingly prominent against blunt end of tail. There are also dark, large webs on the hind toes. Females in terrestrial coloration show a pale line along the middle of their back. Orange streak along the length of the tail is bordered by dark spots on the upper- and underside. Pads of the two outer toes have one pale spot, especially in females.

Distribution and habitat Western Europe, from northern Portugal across France and Britain to central Germany. Especially in forested hillsides up to an altitude of about 1,000 m, in the Pyrenees up to 2,400 m. During breeding season found in small, cool and still water bodies such as ponds, ditches, puddles or flooded areas. In summer found on land in sunny deciduous forests.

Notes Mating takes place between March and early July, in Spain as early as January. 300–450 eggs are singly deposited on the leaves of water plants by the female. Larvae hatch after a period of about 2–3 weeks and reach metamorphosis after 6–7 weeks with a length of 4–5 cm.

Subspecies The nominate subspecies *T. h. helveticus* occurs throughout most of the range, while the status of the subspecies, *T. h. punctillatus* and *T. h. sequeirai*, is still debated.

Similar species Common Newt (p. 38). Bosca's Newt (p. 44) has

Unspotted underside of female Palmate Newt

just one groove on head.

Additional species

■ **Montandon's Newt** *Triturus montandoni* (distribution in blue), which only occurs in the Tatra Mountains and the Carpathians, shows three grooves on the head and a tail filament, but has an orange-red, mostly unspotted underside and no webbing between the toes.

Pair of Palmate Newts

Bosca's Newt
Triturus boscai
Salamanders and newts SALAMANDRIDAE

> ♂ 6–7.5 cm, ♀ 8–9.5 cm
> Pond newts
> Head unstriped; pale band along flanks; ♂ lacks body crest and has no webbing between toes

Description Small, blunt-nosed, smooth-skinned pond newt. Head with one longitudinal groove. Back brownish yellow to olive with irregular black spots. Shows reddish dorsal line in terrestrial coloration. Throat and belly orange-red to yellow with black spots, towards back with pale, diffuse stripes. Male lacks body crest. Male and female have a small tail crest during breeding, and a small tail filament. There is no webbing between the toes.

Distribution and habitat Inhabits montane regions of Portugal and western Spain, also forests and farmland. When breeding found in small, even artificial ponds or slow streams. In summer lives on land under stones, etc.

Notes Breeds December–June. Behaviour as Alpine Newt (see p. 35), but display is more complicated. 150-250 eggs laid on aquatic plants. Metamorphosis takes place in May and June.

Similar species Palmate Newt has no spots on the belly.

Italian Newt
Triturus italicus
Salamanders and newts SALAMANDRIDAE

> 6–8 cm
> Pond newts
> Pale patch behind eye; throat darker than belly;
> ♂ lacks body crest and has no webbing between toes

Description The smallest European newt. Head has one longitudinal groove. Back brownish with dark spots, in terrestrial coloration also has shining golden spots. Underside orange-yellow, spotted black. Throat darker than belly. Longish pale patch behind eye. Male lacks body crest, both sexes have a small tail crest and a short tail filament. There is no webbing between the toes.

Distribution and habitat Found in central and southern Italy from the lowlands up to 1,500 m. During breeding occurs in small, well-vegetated and often temporary ponds. On land found under stones and dead wood.

Notes Breeding from January to May. Eggs deposited singly on aquatic plants in smaller ponds. Metamorphosis takes 4–6 weeks. Neoteny occurs in wells (see Alpine Newt, p. 36).

Similar species Common Newt has a body crest and three longitudinal grooves on head, with throat paler than belly.

Northern Crested Newt
Triturus cristatus
Salamanders and newts SALAMANDRIDAE

> ♂ 10–16 cm, ♀ 11–19 cm
> Pond newts
> Flanks with white stippling; ♂ with high body crest
  which is separate from tail crest; belly brightly
  coloured orange and black

Description Strong, broad-headed pond newt with coarse skin. Upperside dark brown, grey-brown or black-brown with more or less obvious black spots and white stippling on flanks. Terrestrial animals, especially from northern Europe, may be completely black on back. Belly orange-red to yellow with large black spots. Throat dark orange to black with small white spots. During breeding season males have high, spiky body crest, separated from high tail crest by obvious gap. Females lack body crest, but have a low tail crest. Whitish or bluish streak along sides of tail in males. Tail of female is orange below.

Distribution and habitat From Britain and France across large parts of northern and central Europe east to the Urals. Favours lowlands and low hills up to 600 m, rarely occurs up to

46

1,000 m and at the most up to 1,700 m. During breeding found in permanent, stagnant water bodies with dense vegetation, such as ponds and ditches, also in flooded areas. Prefers open habitats (farmland, gravel pits), but also occurs in forests. In summer found on land under dead wood, mostly near water. Spends winter buried in soil.

Notes Until recently the Northern Crested Newt was lumped with the three other European crested newts as a single species (formerly known as Crested Newt *T. cristatus*). Although the forms have been split and are now treated as separate species they can produce fertile hybrids. Mating takes place between March and June. Adults leave the water from July onwards. Females produce a clutch of 200–400 quite large eggs, which can easily be distinguished from eggs of smaller newts by their yellowish-green colour. Eggs are deposited on aquatic plants over a period of several weeks. Larvae hatch after 2–3 weeks. Metamorphosis of the larvae, which can grow up to 8 cm long, takes three more months.

Similar species Other crested newt species can easily be distinguished by distribution. The Alpine Newt in terrestrial coloration is very dark on the upperside, but shows no dark patches on the belly.

Pair of Northern Crested Newts (female with orange belly)

Additional species

■ **Danube Crested Newt** *Triturus dobrogicus* (distribution in blue) occurs only on the central and lower Danube plains. At 12–13 cm long it is a little smaller and more gracile than other crested newts, but it has the highest and proportionally most intensively jagged crest during breeding. It is reddish-brown to brown above with the flanks stippled white. The belly is intensely orange-red, with dark spots that often merge to form longitudinal bands.

Danube Crested Newt

Italian Crested Newt
Triturus carnifex
Salamanders and newts SALAMANDRIDAE

> ♂ 10–15 cm, ♀ 12–18 cm
> Pond newts
> Flanks with dark, round patches and no white stippling; ♂ with body crest; belly orange-black

Description Strong, broad-headed pond newt with relatively smooth to finely granulated skin. Upperside grey, black-brown or yellowish with obvious large, black spots. Similar to Northern Crested Newt (p. 46), but flanks usually lack white stippling. Young animals and females in terrestrial coloration often have a reddish dorsal line. Belly orange-red with large, black spots, throat dark with small white spots. Males during breeding season have a ragged crest that is a bit lower than that of Northern Crested Newt, but which has an obvious gap between the body crest and tail crest. Females lack body crest but have a low tail crest. Males have a whitish or bluish longitudinal line on tail. Females do not show this line, but have an orange underside to the tail. Terrestrial coloration is usually darker than breeding coloration.

Distribution and habitat From the north-eastern parts of the

Alps in Austria across southern Switzerland (released close to Geneva) to Italy and in the western Balkans. In south-east Bavaria there is a contact zone with Northern Crested Newt, there are no pure Italian Crested Newt populations but hybrids between both species occur. Found at altitudes from 400 m to 1,600 m, the subspecies *T. c. macedonicus* up to 2,140 m. During spring breeding season found in large, permanent, stagnant water bodies or in slow-flowing streams with much underwater vegetation. In summer found on land, in deciduous forests or meadows. **Notes** Breeding is from March to June and the animals leave the water no later than August. Egg laying and development as described under Northern Crested Newt (see p. 46). **Subspecies** Macedonian Crested Newt *T. c. macedonicus* from the Balkans was once treated as a subspecies of the Balkan Crested Newt, but is now classed as a subspecies of Italian Crested Newt. The nominate subspecies inhabits Austria and Italy. **Similar species** Other species of crested newts. **Additional species** ■ **Balkan Crested Newt** *Triturus karelinii* (range in blue) is a large, sturdy species. The nominate race occurs in Turkey and the Caucasus. A recently described subspecies, *T. k. arntzeni* occurs

Female T. c. carnifex

in the south-eastern Balkans (north-east Greece and Bulgaria). It has a grey upperside with large, dark spots and a less high and jagged body crest, which is divided from the tail crest by just a small gap.

Male Macedonian Crested Newt

Marbled Newt
Triturus marmoratus
Salamanders and newts SALAMANDRIDAE

> 12–16 cm
> Pond newts
> Upperside marbled black-green; breeding ♂ has smooth-edged crest which is notched in the middle

Description Relatively large, rough-skinned newt with broad, short head. Upperside bright green with irregular dark patches, which form a conspicuous marbled pattern. Immatures and females in terrestrial coloration often show a reddish central dorsal line. Belly pale grey to dark grey with dark patches and spots. During courtship the male has a high, smooth-edged crest that is striped yellowish and dark brown. The body crest merges into a high tail crest, and the two crests are separated by a small gap. The male's tail has a silvery lateral stripe. The female does not show this feature and has no body crest, but it does have a small tail crest.

Distribution and habitat Occurs in western France and Spain, inhabiting plains and hilly areas up to altitudes of 2,000 m. In the breeding season found in still or slow-flowing stretches of water, such as ponds, small lakes, wells or ditches. In

50

summer found on land, under stones and dead trees. In some locations the species remains in the water year-round.

Notes Although it has a very distinct pattern, Marbled Newt is closely related to the crested newts. Hybrid populations between Marbled Newt and Northern Crested Newt occur in north-west and central France, and these were originally described in error as a separate species *Triturus blasii*. Breeding takes place in spring and summer in northern parts of the range and at higher altitudes, or even in winter in southern regions. The female deposits 200–400 eggs per season, which she folds into the leaves of aquatic plants with her hind legs. Larvae reach a length of 8–9 cm. Marbled Newts can live up to 25 years.

Many females and immatures show a red central dorsal line

Additional species
■ **Southern Marbled Newt**
Triturus pygmaeus (distribution in blue) is very similar, but even smaller and more delicate than Marbled Newt. It lives in the south-west of the Iberian Peninsula and has only recently been split as a separate species.

Adult male Marbled Newt in aquatic breeding coloration

Italian Cave Salamander
Speleomantes italicus
Lungless salamanders PLETHODONTIDAE

> 10–12.5 cm
> European cave salamanders
> Webs on fingers and toes; groove from nose to edge of lip; belly dark; snout with distinct canthi

Description Small, gracile land salamander with large eyes, paratoid glands, obvious snout canthi and nasolabial groove (groove between nostril and upper lip; this can only be seen with a magnifying glass). Back coloration very variable, mostly shining yellowish golden or gleaming red to brownish with dark and light patches of different intensity. Belly dark, with lighter patches. Toes are relatively pointed. The males are often a little smaller than the females and possess chin glands during the breeding season.

Distribution and habitat Found only in parts of north and central Italy, where it can be relatively common (and have high population densities) at altitudes up to 1,600 m. Favours moist and cool caves, old stone walls or crevices under dead wood and flat stones.

Notes Land-living amphibian that swims badly but can climb very skilfully, even on vertical rocks. Cave salamanders have no lungs and they breathe with the surface of their body and the lining of their mouth. To catch their invertebrate prey their tongue, which can be as long as their body, shoots out of their mouth like a slingshot. These cave salamanders are probably active all year round. They mate and lay eggs on land. The male grasps the female on its back and the female takes the spermatophore into its cloacal region (internal fertilization). The clutch consists of 5–15 eggs. They are deposited in underground cavities and crevices in spring and are guarded by the female. Development takes 6–11 months.

Italian Cave Salamander

Additional species Two very similar species occur in south-east France and north-west Italy. Both have paler bellies than the Italian Cave Salamander. They are usually less brightly coloured, but can only be identified with certainty by their distributions.

■ **Strinati's Cave Salamander** *Speleomantes strinatii* (range in green) is brownish above with dark patches. It is found in north-west Italy (west of La Spezia) and south-east France, in moist forests and cool valleys up to 2,000 m.

■ **Ambrosi's Cave Salamander** *Speleomantes ambrosii* (range in blue) occurs only in a small area east of La Spezia.

Strinati's Cave Salamander

Gené's Cave Salamander
Speleomantes genei
Lungless salamanders PLETHODONTIDAE

> 10–12.5 cm
> European cave salamanders
> Webs on fingers and toes, blunt tips; belly pale; nasolabial groove; only found on Sardinia

Description Similar to Italian Cave Salamander (see p. 52). The four Sardinian salamander species are very difficult to tell apart. Compared to mainland species they have no canthi on snout and a pale belly with a few spots. Brown above with yellow patches. Broad, blunt tips to toes.
Distribution and habitat Common in south-west (near Iglesiente, Sulcis) in shadowy, moist places in caves and under stones up to altitude of 1,200 m.
Notes See *S. italicus* (p. 52). Emits an aromatic scent when touched.
Additional species Three slightly larger cave salamander species also occur on Sardinia:
■ **Monte Albo Cave Salamander** *Speleomantes flavus* (range in green) inhabits the north-east of the island near Monte Albo.
■ **Supramontane Cave Salamander** *Speleomantes supramontis* (range in orange) in central east (Gulf of Orosei).
■ **Scented Cave Salamander** *Speleomantes imperialis* (range in blue) in south-east (e.g. Nuoro).

Olm
Proteus anguinus
Olm PROTEIDAE

> 20–25 cm
> Olm
> Eel-shaped; white skin; feathery gills; reduced limbs; eyes very small; aquatic; almost exclusively in caves

Description Long and eel-shaped with spatulate snout and crest-like fins on tail. Skin pinkish to yellowish white, becoming dark when exposed to sunlight. Immatures have diffuse grey patches. Three pairs of red gills and small eyes beneath the skin. Slender limbs; front limbs with three toes, hind limbs with two.

Distribution and habitat Lives in underground streams in caves in limestone karst country along the coast from north-east Italy to Montenegro. After heavy rainfall olms can be washed out of caves and into open streams.

Notes Olms are neotenic (see Alpine Newt, p. 36), albinotic permanent larvae. Egg clutches are laid on the surface of stones and are guarded by the female. Can live for up to 60 years. Feeds on worms and isopods.

Subspecies A Slovenian pigmented form with eyes was described as *P. a. parkelj*. It is possible that geographically separated populations represent different species.

55

Fire-bellied Toad
Bombina bombina
Fire-bellied toads BOMBINATORIDAE

> 3–5 cm
> Fire-bellied toads
> Belly orange-red with black patches which enclose
  whitish spots; inner toes dark grey; toe tips blackish

Description Dorsally flattened body and blunt snout. Pupils heart-shaped, eardrums not visible, no parotoid glands. Skin of back covered with with flat, soft warts each of which has a small black spot in the centre. Upperside dark grey to pale grey, brownish or greenish, sometimes with green patches or a green central line. Belly bright orange to red with pattern of black or grey spots. Compared to the similar Yellow-bellied Toad (p. 58) the dark patches always enclose small white spots and the tips of all toes and entire inner toes are dark grey to black rather than yellow. Has obvious webs between hind toes. Males have inner vocal sac and nuptial pads during courtship.

Distribution and habitat Found in eastern Europe, from eastern Germany and Austria eastwards and from Croatia across the eastern Balkans to the Black Sea. Typical lowland species, mostly found at altitudes below 200 m, rarely reaching 600 m.

The Fire-bellied Toad favours warm temperatures and is highly aquatic. It prefers sun-exposed, well-vegetated, still areas of permanent water, such as ponds, ditches or flooded areas. It hibernates in the ground on land close to its breeding site.

Notes Fire-belled Toads are active during both day and night from March to October. Their diet mainly consists of insects and spiders. Breeding takes place between May and July; their loud and melodic *oop oop oop* calls carry a long way. Defence mechanism as described under the closely related Yellow-bellied Toad (see p. 60). The two species may hybridize where ranges overlap. During courtship, the males grasp the females around the loins. Altogether 80–300 eggs are laid in small clumps of 20–30 eggs among water plants. The larvae reach a length

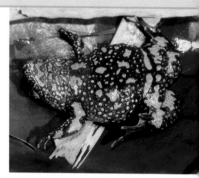

Underside with typical pattern

of 5 cm and metamorphose after 2–3 months. The red colour of the belly can already be seen in immatures which are just 1–1.5 cm long. They remain close to the breeding site but may wander, especially after heavy rain.

Similar species Yellow-bellied Toad is distinguished by the yellow tips of toes, all-yellow inner toes and more extensive bright coloration with isolated black patches on the belly.

Adult Fire-bellied Toad

Yellow-bellied Toad
Bombina variegata
Fire-bellied toads BOMBINATORIDAE

> 3–5.5 cm
> Fire-bellied toads
> Belly yellow with more or less uniform black spots; inner toes yellow; tips of toes yellow

Description Body dorsally flattened and lacks parotoid glands. Skin very warty and rough; in the middle of the larger warts there are often several tiny black spikes. Pupil heart-shaped; no visible eardrums. Upperside yellowish brown, yellowish or grey to olive brown. Belly bright yellow to pale orange with dark grey or black patches. Compared to Fire-bellied Toad (see p. 56) there are no or only very faint white spots on the dark parts of the underside, the tips of the toes are yellow, and the entire inner toes are yellow. The coloration and pattern of the belly is very variable, and almost uniform yellow as well as uniform black individuals (very rare) may be encountered. In contrast to the Fire-bellied Toad, which usually shows no contact zone between the patches of colour on the limbs and those on the belly, these are usually connected on the Yellow-bellied Toad. The pattern of the underside of the Yellow-bellied Toad remains the same throughout its entire life so individuals can be recognized after many years. Has well-developed webs between hind toes. Males lack vocal sacs but during breeding have nuptial pads on the inner sides of the front legs and several fingers.

Distribution and habitat Found throughout much of central and southern Europe, from France across the central German mountains and south to Greece. Absent from the Iberian Peninsula and replaced by the Appenine Yellow-bellied Toad (see p. 61) south of the Po Estuary in Italy. A wide zone of range overlap with the Fire-bellied Toad runs from Germany across the Balkans to Bulgaria. In this area there are

'Unken reflex' defence posture

small but persistent populations of hybrids with fertile individuals which show features of both species. While the Fire-bellied Toad usually occurs in lowlands, its close relative tends to be found in more hilly or mountainous regions. In the north of its range the Yellow-

Pair in amplexus

Underside of a Yellow-bellied Toad from Greece (B. v. scabra)

bellied Toad is most often found at altitudes between 300 m and 800 m, while in the Alps it occurs up to 1,900 m and in the southern Balkans even up to 2,200 m. It is a water- and warmth-loving species, and is often found in smaller, temporary, shallow, sun-exposed and sparsely vegetated stretches of water such as ditches, large puddles, ponds or other similar places. Often occurs in the vicinity of forests, but also in open places created by humans, such as gravel pits

Greek Yellow-bellied Toad

or military training areas, in which there can be a mosaic of many small ponds and ditches. In the south of its range often occurs in areas periodically flooded by mountainous streams. Spends the winter on land close to water.

Notes Yellow-bellied Toads are well known for their defence posture, the so-called 'unken reflex'. If threatened, they strongly arch their back and the arms and legs are bowed sidewards and upwards. In this position, parts of the yellow underside become visible, especially on the limbs. In contrast to popular belief, the animals do not turn around onto their back to present their belly. Another means of defence is a strongly irritating whitish skin secretion. This species is generally active day and night from April–October. Males call *puu puu puu* rather like a Scops Owl. Breeding takes place between May and July, and egg-laying is triggered by a period of regular rainfall. Usually just a few of the females lay eggs. The males grasp the females around their loins (inguinal amplexus). 100–250 eggs are deposited in small clumps of 10–30 onto sub-aquatic vegetation. Larvae hatch after about 2–3 days, and metamorphosis takes place after another 6–9 weeks. Yellow-bellied Toads mature after about two years and can live up to 30 years in captivity.

Subspecies The northern part of the range is occupied by the nominate subspecies, *B. v. variegata*, which shows enormous genetic variation between western and eastern populations. *B. v. scabra* occurs in the southern Balkans (Greece Montenegro, Albany and Bulgaria), while another subspecies described and still under consideration is *B. v. kolombatovici*, which inhabits northern Dalmatia. Both forms can be distinguished from the nominate subspecies genetically and show slightly different patterns. The Dalmatian population in particular shows a tendency towards a plain yellow underside.

Additional species

■ **Appenine Yellow-bellied Toad** *Bombina pachypus* (range in blue) occurs in Italy south of

Mating pair of Appenine Yellow-bellied Toads

the Po Estuary. It was regarded as a subspecies of *B. variegata* until recently. Although there are few distinct features (the throat is often dark and the belly plainer yellow-orange), the genetic differences are so considerable that it is now accepted as a separate species.

Variation in undersides of three Appenine Yellow-bellied Toads

Common Midwife Toad
Alytes obstetricans
Midwife toads/painted frogs DISCOGLOSSIDAE

> 4–5.5 cm
> Midwife toads
> Pupils vertical, slit-like; three tubercles on palms of hands; flanks with row of reddish warts

Description Body small, plump and short-legged, skin warty. On each side of the back there is a row of large, mostly red coloured warts that start on the small parotoid glands and continue back to the base of the hind legs. Vertical, slit-like pupil and visible eardrum. Upperside with variable coloration, mostly grey, brown or olive, sometimes with faint black or greenish spots and patches and often a pale inverted V-shaped patch. Belly whitish, patchy grey on chest and belly. Three roughly equal-sized tubercles on palms of the hands. No spades on feet or webbing between toes. Males lack vocal sacs and nuptial pads and often have strings of eggs wrapped round the hind legs.

Distribution and habitat Found in western Europe, from Portugal to central Germany. Favours forested hillsides, mostly at altitudes between 200 m and 1,000 m, in the Pyrenees up to 2,400 m. Favours moist, wet, sun-exposed sites with sparse vegetation. In north of range often found in farmland, gravel pits and quarries; in Iberia found in a wider range of habitats. Larvae develop in ponds, puddles, ditches or mountain lakes. Hides in walls or under stones, often close to water.

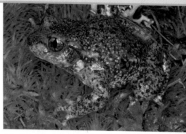

Male Common Midwife Toad of the nominate subspecies

Notes Nocturnal. Breeding generally takes place from March to August. Males give a high-pitched, musical *toop toop toop* call which, from a distance, sounds like the ringing of bells. Mating takes place on land. The male grasps the female around her loins. Strings of 20–80 eggs are connected by a jelly-like substance and wrapped around the male's hind legs. Some may carry up to 170 eggs (the clutches of several females). The larvae are ready to hatch after 3–6 weeks and they are deposited in different stretches of water. They reach a length of 5–10 cm. In cool summers they cannot complete their metamorphosis and have to hibernate as larvae in the water.

Subspecies The nominate *A. o. obstetricans* occurs through most of the range; *A. o. boscai* is in north-west Iberia and *A. o. almogavarii* in north-east Spain. A new subspecies *A. o. pertinax* was recently identified in Albacete, central Spain.

Similar species Other species of midwife toad: palms of *A. cisternasii* have two tubercles, *A. dickhilleni* has no reddish lateral line, *A. muletensis* is gracile and only occurs in north Mallorca. Spadefoots have seams of skin along the hind toe. True toads have a vertical pupil.

Albinotic tadpole just before metamorphosis

Iberian Midwife Toad
Alytes cisternasii
Midwife toads/painted frogs DISCOGLOSSIDAE

> ♂ 3–4 cm, ♀ 3.5–4.5 cm
> Midwife toads
> Pupils vertical, slit-like; palms with two tubercles; fourth finger very blunt and short

Description Body small and plump with warty skin and small parotoid glands. Pupils vertical, upper eyelid with small, reddish warts. Upperside brownish to grey, mostly with dark patches, which may form two indistinct rows. Belly whitish with pale grey patches. Palms with two tubercles of different sizes, fourth finger short and blunt. No spadefoot or palmations.

Distribution and habitat Endemic to the south-west of the Iberian Peninsula, found in low and middle altitudes, rarely reaching 1,800 m. Favours dry, warm, sun-exposed areas such as olive or oak groves, but may also occur in cultivated areas close to streams. Nocturnal, stays hidden during day except during rain.

Notes Breeds in the cool season from September to May. Like the Common Midwife Toad the males carry 20–180 eggs wrapped around their hind legs for 3–4 weeks.

Similar species See Common Midwife Toad (p. 62).

Southern Midwife Toad
Alytes dickhilleni
Midwife toads/painted frogs DISCOGLOSSIDAE

> 4.5–5.5 cm
> Midwife toads
> Pupils vertical, slit-like; palms with three tubercles; skin relatively smooth and lacks reddish warts

Description Body small and plump, very similar to Common Midwife Toad in size and coloration (see p. 62), but skin smoother and flanks lack the row of red warts. Upperside yellowish, brownish or greenish, mostly with small pale and dark spots. Belly whitish. Palms have three tubercles. No webbing on toes, males lack vocal sacs.
Distribution and habitat Endemic to south and south-east Spain, where found at elevations between 500 m–2,300 m.

Inhabits oak and pine forests, but also open areas and rocky hillsides. Hides under stones close to streams in the summer.
Notes This species was first described in 1995. It breeds in April and May and mating is as described for Common Midwife Toad. The males carry the eggs wrapped around their hind limbs for up to two months. Larvae are deposited in streams or wells and grow up to 6–7 cm.
Similar species See Common Midwife Toad (p.62).

65

Mallorcan Midwife Toad
Alytes muletensis
Midwife toads/painted frogs DISCOGLOSSIDAE

> ♂ 3–3.5 cm, ♀ 3–4 cm
> Midwife toads
> Pupils vertical, slit-like; palms with three tubercles; toes long and slender; parotoid glands inconspicuous

Description The smallest and most gracile midwife toad. Has relatively long hind legs, fingers and toes. Skin quite smooth, parotoid glands barely visible and pupils vertical. Upperside brown, yellowish green or grey, with black and olive patches. Palms have three tubercles and there is no webbing on toes.

Distribution and habitat Endemic to the north of Mallorca. Found under stones and in crevices along streams in barely vegetated, narrow canyons of the Sierra de Tramuntana. In summer the larvae can be found in pools of permanent water.

Notes This species was first described on the basis of fossil evidence and was not found alive until 1979. Breeding as in Common Midwife Toad (see p. 62). Mating from late March to July. Males wrap clutches of 7–20 eggs around their legs for 3–4 weeks; these are deposited in streams just before hatching.

Similar species Common Midwife Toad (p.62).

West Iberian Painted Frog
Discoglossus galganoi
Midwife toads/painted frogs DISCOGLOSSIDAE

> ♂ 4–8 cm, ♀ 3.5–7 cm
> Painted frogs
> Pupils drop-like; eardrum not visible; hind feet webbed; only in Iberian Peninsula

Description Skin smooth with a few warts. Pupils drop-shaped, tongue disc-shaped. Upperside very variable; brownish, greenish, grey, yellowish or reddish, either with dark brown patches surrounded by pale borders, with two pale lateral lines and one central line, or with no pattern at all. Underside whitish. Toes smooth underneath and webbed. Best distinguished from other species of painted frogs by distribution. Males have no vocal sacs.

Distribution and habitat Western and central Iberia. Strongly associated with water, common in small, stagnant or slow-flowing bodies of water.
Notes Active both day and night. Breeds throughout the year. Some 5,000–6,000 eggs are deposited singly or in small clumps into the water.
Additional species
■ **East Iberian Painted Frog**
Discoglossus jeanneae (range in blue) favours limestone areas and is best identified by range.

67

Painted Frog
Discoglossus pictus
Midwife toads/painted frogs DISCOGLOSSIDAE

> 5–7 cm
> Painted frogs
> Pupils drop-like; only southern France,
> north-east Spain, Sicily and Malta

Description Resembles West Iberian Painted Frog (see p. 67). Skin smooth with just a few rows of warts. Hind toes webbed. Upperside brownish to reddish, often with dark patches bordered by a pale edge. Often has a few pale stripes, underside pale.

Distribution and habitat Lowland southern France and north-east Spain, Sicily (up to 1,500 m), Malta, Gozo and northern Africa. Adaptable, found in open habitats with still and running water, also close to settlements.

Notes Climbs and jumps well. Lays eggs in small clumps.

Subspecies Nominate subspecies in Sicily and Malta, *D. p. auritus* from Africa introduced to north-east Spain and southern France.

Similar species Other painted frogs (see p. 67 and p. 69).

Additional species

■ **Corsican Painted Frog** *Discoglossus montalentii* (range in blue) is 5–6 cm and lives by forested mountain streams. Plain brown or has dark patches with no pale lines or borders.

Tyrrhenian Painted Frog
Discoglossus sardus
Midwife toads/painted frogs DISCOGLOSSIDAE

> 5–7.5 cm
> Painted frogs
> Pupils drop-like; dark patches on back (without pale borders); only on Corsica and Sardinia and in Tuscany

Description Skin quite smooth, with low warts or rows of warts. Pupils drop-like. Hind feet webbed. Resembles West Iberian Painted Frog (see p. 67), but brownish above with no pale stripes and variable dark patches that have no pale fringes. Underside pale. Very similar to the Corsican Painted Frog, but body more slender, legs a little shorter and snout more pointed. The fourth finger becomes more pointed towards the tip.

Distribution and habitat Found in lowland and mountain areas (up to 1,800 m) in Sardinia, Corsica and adjacent islands (e.g. Giglio) and Tuscany (Monte Argentario). Close to small, still or flowing bodies of water. Unlike Corsican Painted Frog it may also live in brackish water.
Notes Active day and night.
Similar species Other species of painted frogs. Water frogs have round pupils; parsley frogs have vertical, slit-like pupils.

69

Common Spadefoot
Pelobates fuscus
Spadefoots PELOBATIDAE

> ♂ 4–6.5 cm, ♀ 5–8 cm
> European spadefoots
> Pupils vertical, slit-like; dome-shaped head;
>   pale-coloured spade on hind foot

Description Plump-bodied toad-like shape. Delicate skin, moist and quite smooth with flat warts. Large head with rounded snout and obvious, well-marked dome on top of head. Large, protruding eyes with vertical, slit-like pupils; no parotoid gland, eardrums not visible. Upperside variably patterned, usually pale brown to yellowish brown in males, grey brown to pale grey in females, with a relatively symmetrical pattern of large, olive-brown to dark brown patches or rows of patches. Flanks often show brick red patches and small spots. Belly whitish, dark grey patches. Males lack tubercles on palms. Metatarsal tubercles on hind foot form a strong pale brown spade. Well-developed webbing between the hind toes.

Distribution and habitat Occurs across large areas of central and eastern Europe from eastern France to the Urals. Mostly in lowlands below 200 m, but in the Czech Republic up to 700 m.

Lays eggs in nutrient-rich, well-vegetated water bodies such as ponds, ditches, flooded meadows and lakes. On land, it prefers sandy, light and sparsely vegetated areas, and is even found in cultivated areas. Usually remains hidden during the day. Spends the winter, from October to March, deep underground. Digs by moving hind legs sideways, and has special groups of muscles that help raise the sharp-edged spades.

Notes Skin covered with granules which emit a secretion that smells like garlic. When threatened puffs up its body and utters calls like a screaming baby. May jump at or bite predators such as snakes and herons. Outside the breeding season mostly nocturnal; may be diurnal in the water between March and June. During mating males wrap their arms around the loins of the females. Spawn is quite thick, 20–80 cm-long rows of eggs, which are irregularly ordered. These egg strings are deposited between water plants as the pair swims around underwater vegetation. Between 1,000 and 3,000 eggs are deposited. Depending on the water temperature the larvae hatch after 4–10 days and leave the water after another 2.5–5 months. The tadpoles reach an average length of about 10 cm – much larger than the young spadefoots,

Common Spadefoot. Note the distinctive dome on the head

which are just 2–4 cm long. Tadpoles up to 18 cm long are often found; exceptionally they may reach 20 cm. These spend the winter in water.

Subspecies The North Italian Spadefoot *P. f. insubricus* is found only in the Po Estuary, while the nominate subspecies inhabits the rest of the large range.

Similar species Western Spadefoot (see p. 72).

Individual from Hungary with a typical pattern of dark patches

Western Spadefoot
Pelobates cultripes
Spadefoots PELOBATIDAE

> ♂ 6–9 cm, ♀ 7–10 cm
> European spadefoots
> Pupils vertical, slit-like; head not dome shaped; black spade on hind foot; Iberia and south and west France

Description Similar to Common Spadefoot (see p. 70). Plump body with short hind legs. Skin delicate and quite smooth, with a few flat granules. Strong head has a short, rounded snout. Protruding eyes have vertical, slit-like pupils. No dome on head and no visible eardrums or parotoid glands. Upperside variable, usually brown, yellowish, grey or whitish, with dark olive to dark brown patches, sometimes with symmetrical pattern of longitudinal patches. Belly whitish, sometimes with dark grey patches. Males have no tubercles on feet. Black, strong, sharp-edged spade at the base of the first hind toe. Has well-developed webbing on hind feet.

Distribution and habitat Iberia and parts of France. Especially in lowlands, but also on plains at higher elevations of up to 1,400 m. Favours open areas with light, sandy soils, such as pastures, dunes or cultivated areas. Lays eggs in well-vegetated ponds, ditches and slow-flowing streams. During the day hides under stones or in a burrow which it digs itself. Spends winter in a burrow which it also digs. **Notes** Generally nocturnal; may be diurnal when breeding. Defence behaviour resembles that of a Common Spadefoot. It utters alarm calls and inflates its body like a balloon. Mating takes place between October and May and is triggered by heavy rainfall. The males attract a female with repeated deep monotonous *co co co* calls and wrap their limbs around her loins. The clutches of up to 7,000 irregularly ordered eggs in thick lines of a jelly-like substance are deposited in underwater vegetation. Depending on the water temperature the larvae hatch after 1.5– 2 weeks. They grow to 10–12 cm long, and leave the

Detail of spade on a hind foot

water after another 4–6 weeks. The tadpoles often perish due to the drying out of water bodies. **Similar species** Common Spadefoot has a helmet-like bulge on head. Eastern Spadefoot has a pale brown spade. Midwife toads and parsley frogs lack webbing. Typical toads have a horizontal pupil.

At up to 12 cm long, the tadpole is longer than the adult

Eastern Spadefoot
Pelobates syriacus
Spadefoots PELOBATIDAE

> 8–9 cm
> European spadefoots
> Pupils vertical and slit-like; head not dome shaped;
> pale brown spade; only in southern Balkans

Description Similar to Western Spadefoot (see p. 72). Has vertical slit-like pupils and lacks dome on head. Skin smooth with just a few, often reddish spiky warts. Upperside brown, grey or white with greenish or dark brown patches. Belly whitish and unspotted. Prominent, pale brown spade for digging. Webs of hind toes more indented than in other spadefoots.

Distribution and habitat South-west Asia and southern Balkans in open, steppe-like landscapes and open woods. Lays eggs in deep ponds, ditches or flooded pastures with little vegetation. During the day hides under stones or in a burrow that it digs itself. Winters in a burrow.

Notes Defence behaviour and breeding as Common Spadefoot (see p. 70). Mating takes place after rainfall in February–March. Lays 2,000–4000 eggs, larvae reach 10 cm in length, rarely up to 20 cm.

Similar species See Western Spadefoot (p.72).

Parsley Frog
Pelodytes punctatus
Parsley frogs PELODYTIDAE

> ♂ 3–3.5 cm, ♀ 3.5–5 cm
> Parsley frogs
> Pupils vertical; thin flange of webbing on each hind toe; warty, often green, spotted skin

Description Slender body with long limbs. Glandular fold behind the eye, pupils vertical and skin warty. Upperside brownish, grey or olive, mostly with small black and bright green patches. Belly white. Hind toes have just a thin flange of webbing on each toe. Males have inner eardrums.

Distribution and habitat North Spain, France and north-west Italy, in lowlands and at altitudes of up to 1,500 m. Terrestrial outside breeding season and found in open countryside (even dry cultivated areas) or in light woods, especially close to water. Hides under stones in daytime.

Notes Nocturnal. Jumps and climbs well. Call a Corncrake-like *crek crek*. Mating from October–May. Egg strings 3–20 cm long with 1,000–1,500 eggs.

Additional species

■ **Iberian Parsley Frog**
Pelodytes ibericus (range in blue) inhabits south-west Iberia only. It has shorter legs and more pronounced tubercles at the base of each finger.

75

Common Toad
Bufo bufo
Typical toads BUFONIDAE

> ♂ 6–10 cm, ♀ 7–18 cm
> Typical toads
> Pupils horizontal; paired tubercles on the undersides of hind toes; parotoid glands oblique

Description Strong and stout toad with great size variation. In northern and central Europe males grow to 8–9 cm and females reach a length of 11–12 cm, but most individuals are smaller. In southern Europe males can be larger than 10 cm and females up to 18 cm. Has very warty skin, conspicuous eardrums and large parotoid glands that are oblique if seen from above. Pupils horizontal and oval with copper-coloured eyes. Upperside brownish, yellowish, reddish, grey or olive; uniform or with inconspicuous slightly darker or paler patches. Around the Mediterranean more frequently orange or reddish. Belly whitish with grey patches. Males without vocal sacs, but with strong front limbs and black nuptial pads on the inner sides of the three inner fingers during the breeding season. Some webbing. Paired tubercles on the underside of the longest (median) hind toe.

Distribution and habitat Found across most of Europe, except for some islands, such as Ireland, Corsica, Sardinia, Malta, Crete and the Balearic Islands. In Finland even lives north of the Arctic Circle. Found in lowland as well as mountain areas, in the Alps up to 2,200 m and in the Pyrenees up to 2,600 m. Very adaptive species, widely distributed and very common in some areas. Prefers sunny places for breeding, favouring still and usually large and deep bodies of water, such as lakes and ponds, but also in rivers, streams or rarely in small and temporary puddles. There must be structures such as aquatic plants or overhanging branches

Pair with string of eggs

to which egg strings can be attached. On land favours woods, but also in meadows, gardens, parks and cultivated areas. During the day hides under dead wood or stones. Spends the winter in a burrow.

Notes Common Toads are among the best known amphibians due to their widespread

During the mating season, males will clutch onto almost anything, in this case an Edible Frog

Common Toad of the nominate race with reddish coloration

distribution and common status. Diet: insects, spiders and other arthropods as well as worms and slugs. When faced with a predator, Common Toads lower their head and raise their hindquarters in a threat display. They can produce a copious white secretion from their skin and parotoid glands. Common predators are mustelids, crows and grass snakes, as well as some kinds of flesh flies. These flies lay eggs into the head of the toad and the larvae burrow to the inside of the nose where

Common Toad in water among floating vegetation

they destroy the mucous membrane, bone and eventually the brain, which leads to death. Common Toads often return to their place of birth to breed, although many pairs occupy newly created ponds. Common Toads wander widely throughout the year between breeding areas and summer- and winterquarters. During warm, rainy nights in the breeding season, (in Britain from March to April) toads migrate in large numbers to their breeding sites. During this time, they are active night and day. The males are very enthusiastic. They wrap their limbs around anything that is soft and malleable (and thus resembles a female), for example other amphibians, fish, dead animals or decaying parts of plants floating in the water. Sometimes balls with a single female and up to 15 males occur, and the female often drowns. Usually the male grasps the female just behind the front limbs and in that position the female deposits 3,000–8,000 eggs in a 3–5 m long string under the water. The eggs form a double row inside the string. The black tadpoles hatch after 2–3 weeks and can form huge swarms. They have a bitter substance in their skin which deters predators such as fish, but not the larvae of certain insects, such as dragonflies. After another 2–3 months the tadpoles reach 3–4 cm and metamorphose. The young toads are 7–10 mm long and feed mainly on isopods and small insects.

Subspecies Three subspecies have been described but their status is uncertain as there is no evidence for genetic difference. The subspecies occurring around the Mediterranean (*B. b. spinosus*) is 15–18 cm long and much stronger and stouter than the nominate subspecies which ranges across most of northern and central Europe. Additionally, *B. b. spinosus* has larger parotoid glands and very warty skin with fine, black spikes, which may be an adaptation to the dry conditions where they live. *B. b. gredosicola*, which occurs in the Sierra de Gredos in central Spain, can be distinguished by its smaller size, smoother skin

Common Toad from the Spanish Sierra de Gredos

and contrasting dark brown and yellowish green upperside.

Similar species Natterjack usually has a bright yellow central dorsal line and parallel parotoid glands. Green Toad has bright green patches on the body and small, unpaired tubercles under the toes. Parsley frogs, spadefoots and midwife toads have vertical, slit-like pupils.

Pair of Common Toads from the Mediterranean in amplexus

Natterjack
Bufo calamita
Typical toads BUFONIDAE

> ♂ 4–8 cm, ♀ 5–9 cm
> Typical toads
> Yellow stripe down the centre of back; underside of toes with pairs of tubercles; parotoid glands parallel

Description Plump body. Pupil horizontal to round, iris yellow to greenish, eardrum inconspicuous. Warty, dry skin, parallel, flat, triangular parotoid glands. Upperside brown, grey, yellowish, olive or greenish; mostly with variable dark patches and reddish warts. Usually there is a distinct yellow line that runs from the head along the back. Belly whitish with grey patches and there are granules on the last third of the body. Males have a vocal sac on their throat and, during the breeding season, black nuptial pads on the inner sides of the three inner fingers. Underside of toes with pairs of tubercles; moderately developed webbing between the toes.

Distribution and habitat In western and central Europe as well as some parts of eastern Europe, especially in lowland and hilly areas. In Spain up to an altitude of 2,500 m. Very localized in Britain, where it occurs especially on heathland and in dunes. Generally lives in open, dry and warm places, with sand or other light soil that is easy to dig into. Lays eggs in shallow, sunny and sparsely vegetated bodies of water, often new and temporary puddles.

Notes Rarely walks or hops, but tends to run like a mouse. Males call with a unique, rattling voice, which resembles the voice of a mole cricket, but much louder. Wanders a lot and seldom establishes site fidelity. Relatively long breeding period; in Britain from March–September; in Iberia from December–June. Egg-laying is triggered by rainfall and 2,000–4,000 eggs are deposited in 1–2 m long strings in shallow water. The black larvae hatch after 2–12 days, depending on the water temperature, and develop relatively quickly. They are robust and tolerate salty water and high water temperatures. As

Calling male Natterjack with an inflated vocal sac

with the tadpoles of Green Toad, those of the Natterjack form honeycomb-like patterns on the lake or pond bed (see p. 83). They usually metamorphose within 1–2 months. The diurnal and freshly metamorphosed young toads are just 6–10mm long.

Similar species Other species of toads (see Common Toad, p. 76), spadefoots, midwife toads and parsley frogs.

Male Natterjack from Spain

Green Toad
Bufo viridis
Typical toads BUFONIDAE

> ♂ 6–9 cm, ♀ 7–10 cm
> Typical toads
> Obvious green markings on upperside; single tubercles on undersides of toes; parotoid glands parallel

Description Robust build, pupil horizontal to round, iris green to yellow, eardrums obvious. Skin dry with obvious warts, parotoid glands parallel, also has small groups of glands on thigh and forearm. Upperside pale grey, whitish, brownish or greenish with a clear pattern of large, well-defined dark green to pale green patches and reddish warts. Rarely has white (but not yellow) central dorsal line. Belly whitish to grey, often with small, dark grey to dark green patches. Males with vocal sac on throat and black nuptial pads on the inner sides of the first three fingers during mating season. Small, single, unpaired tubercles on the underside of the longest toe. Moderately developed webbing between the toes.

Distribution and habitat Occurs in eastern and south-eastern Europe, from Germany and Italy east to central Asia; also north Africa and large Mediterranean Islands such as Sardinia, Sicily

82

and Crete; introduced to the Balearic Islands. In the north found in lowlands at altitudes below 500 m. In southern Europe up to 2,400 m and outside Europe even higher. In central Europe very localized, declining in many parts of its range. Typical species of arid steppe; favours open and dry landscapes with light soil, for example, fields, dunes, dry meadows and fallow land. Breeds in stagnant bodies of water with shallow banks, such as ponds and ditches; also tolerates brackish water. **Notes** In Asia there are populations which require further study and couldbe reclassified as a separate species. Diet consists of spiders, mites, ants, beetles and other insects. Mobile and agile species which moves with fast hops. In northern and central Europe active between March and October, especially at night, but may be diurnal during breeding between April to June. The males utter a melodic trill in shallow water. The male wraps its arms around the shoulders of the female and she lays 2–4 m-long egg-strings which contain 2,000–10,000 eggs in shallow water. Larvae hatch after one week and leave a honeycomb-like pattern on the soft ground. This pattern is caused by groups of tadpoles gathering and forming small grooves that gradually

Pair of Green Toads in amplexus

become deeper. After 2–3 months the 10–15 mm-long, diurnal young toads leave the water. **Similar species** See Common Toad (see p. 76).

Tadpoles often form 'honeycomb' patterns on the lake or pond bed

Tree frogs

Common Tree Frog
Hyla arborea
Tree frogs HYLIDAE

> ♂ 3–5 cm, ♀ 4–6 cm
> Tree frogs
> Toes have disc-shaped pads; back green with dark
  stripe running along flanks

Description Small, smooth-skinned, long-legged tree frog with disc-shaped adhesive pads on the toe-tips and fingertips. Pupils horizontal and oval. Eardrums small but obvious. Upperside usually uniformly coloured, pale to dark green (like glossy paint), rarely yellow or brownish, blue, grey or mottled. Individuals can change colour. Shows dark stripe on the sides that runs from the nose across the eardrums to the groin, where it bends upwards and forms a dark patch. Belly whitish cream to pale grey and granular. Toes webbed. Throat white in female, yellowish brown and often wrinkled in male, which inflates a large vocal sac when calling.
Distribution and habitat Widely distributed in Europe, but not in Britain, Scandinavia or much of Italy or southern Iberia. Mainly in lowland and hilly countryside below 800 m, sometimes higher, in Spain up to 2,000 m, in Bulgaria up to 2,300 m. Outside breeding season found on trees,

bushes and reeds, often in sunny places. In breeding season in quiet, sunny ponds, ditches and flooded areas. Hibernates in crevices, trees or underground.

Notes Declining in the north of its range and threatened in many areas. During the breeding season (April to late June) males give loud, rattling calls (rather like a quacking duck) mainly at night from the banks of ponds or while floating on the water. Males grasp the females around their shoulders and they deposit 400–1,400 eggs in small clumps of 10–50 eggs onto aquatic vegetation. The larvae, which favour warm temperatures, have a deep tail fin that starts at the front of the body and a very round, golden- coloured belly. After 8–11 weeks they metamorphose when they are

Newly metamorphosed juvenile

about 5 cm long. Tree frogs can live for up to 15 years.

Subspecies The formerly recognized Iberian subspecies *H. a. molleri* and Cretan sub-species *H. a kretensi*, which has a dark loop around the groin, are no longer regarded as valid.

Similar species Stripeless Tree Frog (see p. 86).

Calling male with inflated vocal sac

Stripeless Tree Frog
Hyla meridionalis
Tree frogs HYLIDAE

> ♂ 4–5, ♀ 4–6.5 cm
> Tree frogs
> Fingers and toes with disc-shaped pads; back
>   grass-green without a stripe on the flanks

Description Long-legged tree frog with relatively short, rounded head, disc-shaped tips of fingers and toes and smooth skin. Obvious eardrums, pupils oval and horizontal. Upperside mostly bright green, rarely spotted, brownish, blue or grey. No dark patch around hind limbs, only a short, dark stripe that reaches from the nostrils across the eardrums to the front limbs. Belly white to pale grey with a granular texture. Insides of the thighs yellowish. Has webbing between the hind toes. Throat whitish green in females, yellowish brown and wrinkled in male. When calling inflates a large vocal sac on its throat which is greenish at the fringes.
Distribution and habitat South-west and north-east Iberia, southern France and north-west Italy, as well as the Balearic Islands, Canary Islands and north-west Africa. Found in lowlands up to 650 m. During the breeding season it favours small, quiet, sunny and well-

86

vegetated bodies of water such as ponds, wells and ditches. Outside the breeding season found in gardens and pastures on bushes (particularly bramble), trees and reeds during the day, often in sunny places. Active all year, seldom hibernates.

Notes Diet: small insects and spiders. Active especially at night and dusk, sometimes during the day. Breeds December– May. The female deposits 1,000 eggs in small clumps of 10–30 onto the leaves of sub-aquatic vegetation. Larvae hatch after a few days and leave the water between June and July. Like the larvae of all other species of tree frogs, they have tall tail fins and

Stripeless Tree Frog in typical pose

a round, golden-coloured belly.

Similar species Common and Italian Tree Frog have an obvious stripe on the flanks and a dark patch around the hind limbs. Tyrrhenian Tree Frog has an indistinct and irregular stripe on the flanks.

Male with yellowish throat

Tyrrhenian Tree Frog
Hyla sarda
Tree frogs HYLIDAE

> ♂ 3–4 cm, ♀ 3–4.5 cm
> Tree frogs
> Fingers and toes with disc-shaped pads; indistinct dark line on flanks; no dark area around hind limbs

Description Smallest of the four European tree frogs, with short snout and relatively short hind limbs. Skin often slightly granular. Fingers and toes have disc-shaped pads. Pupils horizontal and elliptical; eardrums obvious. Upperside variable, often not uniform green but diffuse brown, green, yellow, olive or grey shades and with dark green to grey patches; can change colour with temperature. Dark, indistinct stripe on flanks, which reaches from the nostrils across the eardrums and along the sides, but may be interrupted in the middle of the flanks and does not form a dark patch around the hind limbs. Belly whitish. Hind legs often have dark bars. Webbing between hind toes. Throat yellowish brown and wrinkled in males. Calling males have a large vocal sac. Throat whitish in females.
Distribution and habitat Only in Sardinia, Corsica and some small surrounding islands such

as Elba. In lowland and hilly areas at altitudes up to 1,000 m, often close to the coast. Often in reeds around lakes and ponds or close to wells; also common on forest edges and in gardens. Lays eggs in well-vegetated, sunny bodies of water.

Notes Diet: small invertebrates. Active year-round, especially at night and dusk. During egg-laying the male grasps the female's shoulders. Eggs are deposited in small clumps onto sub-aquatic vegetation. Larvae hatch after a few days and leave the water in early summer.

Additional species

■ **Italian Tree Frog** *Hyla intermedia* (distribution in blue) occurs from southern Switzerland (Tessin) to Italy,

Italian Tree Frog from Sicily

including Sicily. It is similar in appearance and calls to Common Tree Frog. The dark patch around the hind limbs is similarly well developed but the eardrums appear larger and the dark streak around the flanks often starts behind the eye.

Tyrrhenian Tree Frog from Corsica

Common Frog
Rana temporaria
Typical frogs RANIDAE

> ♂ 5–11 cm
> Brown frogs
> Dark ear patch; short legs; small, soft metatarsal tubercles; blunt, rather raised snout

Description Strong, short-legged brown frog with blunt and somewhat raised snout if seen from the side. If the hind legs are carefully bent forward along the body, the joint of the heels does not reach the snout. Upperside smooth or slightly granular in texture with two closely spaced dorsolateral folds. During the breeding season females often show pearly granules on their flanks. Brown ear patch with obvious eardrum in the centre. Below this dark patch the upper lip is often contrastingly pale. Upperside variable, pale brown to dark brown, yellowish, olive or reddish, with irregularly patterned black or brown spots. Sometimes with a pale, diffuse central dorsal line. Belly whitish to grey, usually barred or spotted brownish (reddish brown in females). Strong webbing between hind toes. Metatarsal tubercles flat, soft and short (first toe is 2.2–4 times as long as the metatarsal tubercle). Males have two inner vocal sacs in throat, during breeding season they have black nuptial pads on the thumbs, a bluish throat and a soft body due to an accumulation of lymph under the skin.

Distribution and habitat Widely distributed in northern and central Europe, in the north up to the North Cape and in the east to the Ural Mountains, but does not occur throughout most of Spain, Italy and Greece. Found in both lowlands and mountains; especially in cool, shady places, in the Pyrenees up to almost 3,000 m. Very common, especially in open areas such as gardens, moist pastures and arable land, but also in woods.

Male Common Frog with relatively strong markings

Some populations have been declining at an alarming rate recently. Mating occurs in quiet and slow-flowing bodies of water of varying sizes: ponds, pools, swamps, ditches, temporary puddles or sun-exposed shores of lakes. In the summer often stays close to streams or ponds. Hibernates

Common Frog pair in amplexus

Reddish coloured individual

in the water or on land. Common Frogs hibernating under water are sometimes active underneath a layer of ice. **Notes** Adaptive species that feeds on insects and other arthropods. In Britain active between February and October, especially at night, may be

Newly laid clumps of spawn

diurnal during the egg-laying period. Lays its eggs rapidly, and is present at its breeding site for a relatively short period (up to April). After that they disperse to their summer quarters. During these movements they cover mostly short distances of about 1–2 km, but distances of up to 10 km and an altitudinal migration of 500–1,000 m have been recorded. In the south it is quite common for egg-laying to take place into November, while at higher altitudes it only lasts until June. Males have a quiet, growling voice. They grasp their arms around the females' shoulders. Each couple deposits one or, at the most two clumps of spawn, each with 1,000–4,000 eggs, on the bottom of the pond. Often hundreds of pairs gather in favoured

parts of a breeding site. The numerous deposited clumps of spawn usually rise to the surface of the water due to tiny air bubbles, and they form large carpets. Eggs are resistant to low temperatures and can even survive freezing for short periods. Larvae hatch after 2–3 weeks and metamorphose after 2–3 months when they are about 4.5 cm long. The newly metamorphosed frogs are just 10–15 mm long due to the shortening of the tail.

Subspecies The classification of Common Frog subspecies is still under discussion. The nominate subspecies *R. t. temporaria* is found across most of Europe, whereas *R. t. parvipalmata* occurs in western Spain and *R. t. honnorati* in the south-east of the French Alps. The status of *R. t. canigoensis* from the French Pyrenees is unclear. The Grass Frog from the Spanish Pyrenees was once considered a separate species (*R. aragonensis*) but is now regarded as a subspecies of the Common Frog (*R. t. aragonensis*).

Similar species Moor Frog has a pointed snout, a hard, raised metatarsal tubercle and a pale, well-defined streak down the centre of the back. Agile Frog and Italian Agile Frog both have longer legs (the heel joint reaches beyond the snout if the hind legs are carefully stretched out along the body) and a plain underside. Throat darker in Iberian Frog, Greek Stream Frog and Italian Stream Frog have a paler central dorsal line.

Newly metamorphosed Common Frog, approximately 1 cm long

Moor Frog
Rana arvalis
Typical frogs RANIDAE

> 4–7 cm
> Brown frogs
> Dark ear patch; often has bright dorsal stripe; large, hard and semicircular tubercle; pointed snout

Description Small, short-legged and delicate brown frog with pointed snout profile. If the hind legs are carefully bent forward along the body the joints of the heels reaches as far as the eye, but not beyond the tip of the snout. Eardrum is obvious against the dark ear patch. Skin smooth on the back with obvious whitish folds. Upperside variable and contrastingly patterned; brownish, reddish, yellowish or grey-brown and spotted dark. Flanks are often paler brown than the back and most individuals show a broad, bright, dark-bordered dorsal line (although animals without any pattern sometimes occur). During courtship males appear flabby because of lymph under their skin and often show bright blue to violet coloration. Belly whitish, usually without dark patches. Has webbed hind toes.

Large, solid, semicircular, raised metatarsal tubercle (the first toe is less than 2.1 times as long as the metatarsal tubercle). Males with two inner vocal sacs on throat and dark nuptial pads on thumbs in the breeding season.

Distribution and habitat Northeastern central Europe, northern and eastern Europe to Siberia, especially in lowlands up to 600 m, and rarely as high as 1,000 m. Favours bogs, moors, wet pastures, woods near rivers and similar habitats with plenty of water. Breeds in bogs, ponds and ditches with firm and sunny banks. Hibernates on land and in underwater mud.

Notes Generally active from March–October, especially at night, but during the breeding season can be diurnal. Breeds en masse and is present at breeding sites for just a few days or weeks in March and April. Males call quietly, sound is reminiscent of bubbles

Moor Frog with typical pale broad dorsal line

emerging from a submerged bottle. Altogether 800–3,000 eggs are deposited in 1–2 clumps of spawn under water and they are attached to subaquatic vegetation.

Subspecies Balkan Moor Frog *R. a. wolterstorffi* inhabits the southern and eastern parts of the range, whereas the nominate subspecies occurs elsewhere.

Similar species Common Frog.

Many males are bright blue during the mating season

Agile Frog
Rana dalmatina
Typical frogs RANIDAE

> ♂ 4–6.5 cm ♀ 4.5–7.5 cm
> Brown frogs
> Dark ear patch; long legs; long snout;
>   eardrums as large as eyes; throat pale

Description Mid-sized, quite slender brown frog with long legs and a rather pointed snout. If the hind legs are carefully bent forward along the body, the joint of the heel always rests beyond the tip of the snout. Has brown ear patch and large eardrum which is close to the eye and almost as large. Upper side smooth or slightly granular with two obvious dorsolateral folds. Often lacks contrasting colours; pinkish brown (dead-leaf coloration), sand-coloured or isabelline (yellowish brown), rarely shows strongly patterned back. Belly and throat whitish and usually without spots. Mid-sized metatarsal tubercle, the relation of which to the first toe is somewhere between that of the Common Frog and the Moor Frog. (first toe is 1.8–3 times as long as metatarsal tubercle). Male lacks vocal sacs. Has grey nuptial pads during the mating season.
Distribution and habitat Widespread in central and southern

96

Europe except for most of the Iberian Peninsula. Common in lowland areas, especially at altitudes below 400 m. In the Alps up to 1,700 m. Favours light deciduous forests with dense low vegetation or pastures close to rivers, often on the edge of forests and sunny clearings. During breeding season found in quiet bodies of water such as ponds, puddles or flooded woods close to rivers. Hibernates on land, but males may spend the winter in mud on the bottom of ponds. **Notes** Active February–October, especially at night, may be diurnal in the February–April breeding season. Lays eggs very early; in the north of its range often the first frog species at its breeding sites. Males call very quietly underwater. Spawn

Agile Frog showing typical large eardrum

clumps of 500–1,000 eggs are attached to sub-aquatic plants or branches. Larvae hatch after three weeks, reach about 6 cm and leave the water in June or July. Newly metamorphosed frogs are just 1–1.5 cm long. **Similar species** Common Frog.

Long hind legs are a typical feature of the Agile Frog

Italian Agile Frog
Rana latastei
Typical frogs RANIDAE

> ♂ 4–5.5 cm ♀ 4–7 cm
> Brown frogs
> Dark ear patch; eardrums smaller than eyes; throat dark with pale central line

Description Relatively small, delicate brown frog with dorso-lateral folds, long legs and short, rounded snout. If the hind legs are carefully bent forward along the body, the joint of the heels always falls beyond the tip of the snout. Similar to the Agile Frog (see p. 96), but smaller eardrum further from eye, and has grey speckled belly and dark throat with pale central line. Upperside reddish brown or grey to dark brown, and usually plain. Males have no vocal sacs.

Distribution and habitat Northern Italy (Po Estuary), southern Switzerland and northern Croatia. Found in wooded valleys in lowlands up to 350 m. Breeds in ponds and small lakes.
Notes One of Europe's rarest frogs. Active day and night. Lays eggs from January–April. Deposits small clumps of spawn with 100–900 eggs on sub-aquatic vegetation.
Similar species Brown frogs. Italian Stream Frog is stouter with finely patterned throat.

Italian Stream Frog
Rana italica
Typical frogs RANIDAE

> 4–6 cm
> Brown frogs
> Dark ear patch; long legs; throat dark with whitish patches and bright central line; only in Italy

Description Small and quite plump with short snout and long legs. If the hind legs are carefully bent forward along the body, the joint of the heel always fall beyond the tip of the snout. Upperside smooth with dorsolateral folds, brownish, reddish or yellowish, often with pale spots that resemble mould. Belly whitish, throat dark with pale spots and faint central line. Male has inner vocal sacs.
Distribution and habitat Only Italy, from Genoa southwards.

Found in hills and mountains, on cool mountain streams rich in oxygen and near wells up to an altitude of 1,700 m. Strongly associated with water.
Notes Breeds February–April. Small clumps of spawn with 200–1,000 eggs are attached to stones close to stream banks.
Similar species Other brown frogs, especially Common Frog and Agile Frog. Balkan Stream Frog is larger and stouter. Italian Agile Frog is more delicate with coarser patterns.

Balkan Stream Frog
Rana graeca
Typical frogs RANIDAE

> 5–8 cm
> Brown frogs
> Dark ear patch; throat dark with whitish patches and pale central line; only south-west Balkans

Description Mid-sized, plump, long-legged brown frog with short snout. If the hind legs are carefully bent forward along the body, the joint of the heels always falls beyond the tip of the snout. Upperside smooth with dorsolateral folds, brownish, reddish, yellowish or grey to olive brown, often spotted dark and with white spots resembling mould. Underside pale, throat dark with pale spots and pale central line. Males have inner vocal sacs and nuptial pads.

Distribution and habitat South-western Balkans, hilly country, mountain ranges 100–2,300 m. Cool flowing streams, mountain streams without vegetation, or slow-flowing parts of rivers.
Notes Often sits on stream banks, jumps into water when in danger, hides under stones. Breeds February–April. Clumps of spawn with 200–2,000 eggs are deposited under stones.
Similar species Other brown frogs. Agile Frog has plain throat. Italian Stream Frog smaller.

Iberian Frog
Rana iberica
Typical frogs RANIDAE

> ♂ 3–5 cm ♀ 4–7 cm
> Brown frogs
> Dark ear patch; long legs; throat with dark barring and bright central line; only on Iberian Peninsula

Description Delicate brown frog with short, rounded snout and long legs. If the hind legs are carefully bent forward along the body, the joint of the heels always rests beyond the tip of the nose. Brown ear patch. Upper side smooth with widely spaced dorsolateral folds, brown, dark olive or grey, often with black and white spots that resemble mould. Belly whitish with dark patches, throat has dark barring and a bright central line. Males have no vocal sacs.

Distribution and habitat Endemic to north-west Iberia, where found in cool mountain forest streams between 100–2,300 m.
Notes Often rests on the banks of streams and jumps into the water when in danger. Mating season depends on the altitude, usually November–May. Small clutches of 500 eggs laid under stones.
Additional species
■ **Pyrenean Frog** *Rana pyrenaica* (range in blue) is very similar and was first described in 1993. It occurs only in Pyrenean streams.

Pool Frog
Rana lessonae
Typical frogs RANIDAE

> ♂ 4.5–6.5 cm, ♀ 5–7.5 cm
> Water frogs
> Large, semicircular, metatarsal tubercles; lower part of hind leg short; back of thighs bright yellow

Description Generally smaller than Marsh Frog; quite slender with a pointed snout and short lower legs. Upperside variable, mostly grass-green or bronze, sometimes bluish green, and with distinctive roundish dark brown patches and occasionally with a yellowish to green central dorsal line. Obvious dorsolateral folds, limbs brownish on the upperside. During the mating season the males acquire a lemon yellow coloration on the head and back. They develop two white lateral vocal sacs and grey nuptial pads on the thumbs. Back of thighs strongly yellow to orange with obvious barring. Underside white with a few grey patches. Well-developed webbing on hind limbs. The best identification feature for the three widespread species of water frog is the inner metatarsal tubercle on the first front toe, which is large, sharp-edged and semicircular in the Pool Frog and reaches almost two-thirds of the length of the first toe (the first toe is less than 2.1 times as long as the metatarsal tubercle).

Distribution and habitat Found in most of central and eastern Europe west to western France and south to the River Po, with scattered northerly populations as far north as southern Sweden. Favours lowland areas, but also found in hilly regions; in the Alps up to 1,500 m. Prefers open habitats close to small, permanent, stagnant well vegetated bodies of water, e.g. bogs in forests, ponds, pools and ditches, but also larger pools in forests with well vegetated and sun exposed areas. Rarely found near human habitation. Hibernates on land, occasionally under water. In general, a little less associated with water than Marsh and Edible Frogs.

Notes Active March–October, night and day. Males call with a loud rasping *kaw-wack kaw-wack* in breeding season from late April to early July. A total of 400–4,500 eggs are deposited in small clumps of spawn on sub-

The yellow-and-black barred hind limbs are obvious when swimming

aquatic vegetation in shallow water. Depending on the temperature, larvae hatch after a few days and metamorphose after 2–4 months. Diet: spiders, insects, slugs and worms. Main predators are mammals such as mustelids and foxes; birds such as storks and herons; grass snakes and fish.

Similar species Marsh and Edible Frogs are distinguished by their smaller metatarsal tubercles, longer lower legs and less regular, rasping voice, which is more cackling in the Marsh Frog.

Pool Frogs often rest on the leaves of Yellow Water-lilies

Edible Frog
Rana kl. *esculenta*
Typical frogs RANIDAE

> ♂ 5.5–10 cm ♀ 6.5–12 cm
> Water frogs
> Metatarsal tubercle raised, but not semicircular; back of thighs mostly only with a hint of yellow

Description A stable hybrid between Pool Frog and Marsh Frog, which is variable and shares characteristics of both species. Coloration similar to Pool Frog (see p. 102), but males have pale grey lateral vocal sacs and yellow tones less intense or lacking. The backs of the thighs are only weakly barred yellow and brown. Back is grass-green to pale green or brownish with dark patches and often with a pale central dorsal line. Prominent, often bronze-coloured, dorsolateral fold. Upper side of limbs brownish. Belly whitish, often with grey patches. Strong webbing on hind limbs. Inner metatarsal tubercle quite large and not semicircular; highest point is towards the toe-tip. Metatarsal tubercles about half to two-thirds of the length of the first toe (first toe between 1.9 and 2.6 times length of tubercle). **Distribution and habitat** Almost identical to that of the Pool Frog. Occurs across much of northern and central Europe. Introduced in parts of Britain. Found in lowlands and hilly areas up to 1,500 m in several different habitats, and is usually is the most common water frog. Prefers small ponds, but also inhabits larger, quiet bodies of water, such as lakes and pools with sunny, well-vegetated banks. Hibernates under water or on land.

Calling male with pale grey inflated lateral vocal sacs

Notes The Edible Frog is a hybrid between Pool Frog and Marsh Frog. Half the set of chromosomes is from each parent species. It is able to reproduce thanks to a mechanism called hybridogenesis, which leads to the elimination of one of the half sets of chromosomes during the process of the formation of eggs or sperm. This usually happens to the chromosomes that belong to the the Pool Frog or Marsh Frog if they mate with an Edible Frog. For that reason, mating between a hybrid and one of the original species always results in hybrids. In order to clarify this situation and to distinguish between the original species and regular hybrids, these special hybrid forms are called kleptons, (from the Greek word 'kleptein', meaning 'to steal') as is reflected in the name *Rana* kl. *esculenta*. The young from matings between Edible Frogs are usually infertile, but in some places stable hybrid populations are known in which individuals have an extra set of Pool Frog chromosomes (triploid individuals).

Similar species Other species of water frogs.

Eggs are laid in small clutches

Marsh Frog
Rana ridibunda
Typical frogs RANIDAE

> ♂ 5–10 cm ♀ 6–14 cm
> Water frogs
> Metatarsal tubercles small, flat, triangular;
  backs of thighs lack yellow tones; dark vocal sacs

Description Large, strong water frog with long legs and blunt snout. Skin quite warty and rough, with obvious dorsolateral folds. Upperside mostly olive-brown to grey, rarely dark green to pale green or yellowish with irregularly formed dark patches and usually with a bright central dorsal line. Underside whitish with dark patches, may be white in south-eastern parts of its range. Backs of thighs whitish or grey with dark barring and no yellow tones. Males have dark grey lateral vocal sacs and grey nuptial pads during breeding. Strong webbing. Inner metatarsal tubercle small, triangular and flat (first toe is more than 2.4 times the length of the tubercle).

Distribution and habitat Across much of Europe. Introduced to south-east England and Switzerland. Usually in lowlands, but also in low mountains, in the south up to 2,000 m. Aquatic, gregarious species that favours open areas close to quiet or slow-flowing bodies of water, often on lakes with sunny, well-vegetated banks. Inhabits rivers across its distribution, but can be found in very small temporary ponds. Hibernates in water.

Notes The correct taxonomy of water frogs is still widely debated because many species are still often lumped under *R. ridibunda*. Latest results of intense research indicate that the central and southern European populations belong to the same species, but that they are genetically quite different from populations in eastern Europe. If these findings are confirmed Marsh Frog could be split into more than one species. Marsh Frogs are active from April to October, and breeding mainly takes place in May and June. Up to 10,000 yellowish eggs are deposited in numerous clumps on sub-aquatic vegetation. Larvae grow to between 7–9 cm long. During breeding has a crackling call

Individual with brown coloration

made up of bursts of 'quacking' croaks.

Similar species Pool Frog and Edible Frog both have larger and more rounded metatarsal tubercles and less crackling and more rattling voices.

■ **Greek Marsh Frog** *Rana balcanica* (range not marked on map) replaces Marsh Frog in Greece and Albania. It is generally browner with a more prominent pale dorsal stripe and grey-black vocal sacs.

Greek Marsh Frog, northern Greece

Epirus Water Frog
Rana epeirotica
Typical frogs RANIDAE

> ♂ 6–8.5 cm, ♀ 6.5–10 cm
> Water frogs
> Metatarsal tubercle very small and triangular; thighs green with dark, bow-shaped bars

Description Mid-sized water frog with low dorsolateral folds and small, low, triangular metatarsal tubercles. Upperside olive to pale green with large, irregular, dark green patches and often a pale green central dorsal line. Underside whitish. Strong, yellowish webbing on hind toes. Upper side of thighs green with bow-shaped, dark barring, which reaches down to the underside. During breeding males have dark grey to black lateral vocal sacs and dark nuptial pads.

Distribution and habitat In west of Greece west of the Pindos mountains, including Corfu and parts of the Peloponnese, as well as southern Albania from sea level up to 500 m. Found in lakes or ponds and on quiet parts of rivers and channels.

Notes The females deposit about 2,000–3,000 eggs in March or April. May hybridize with Greek Marsh Frog where both species occur together.

Additional species In the southern Balkans and Greece

there are four additional species of water frog (not including the Greek Marsh Frog *R. balcanica*, see p. 107, which is sometimes treated as *Rana kurtmuelleri* by some experts). These four species are very difficult to tell apart and do not usually occur together, although they may hybridize where their ranges do overlap.

The taxonomy of water frogs in Greece is still unclear

■ **Albanian Pool Frog** *Rana shqiperica* (range in green) is 6–8 cm long and can be found in the coastal lowlands of Albania and Montenegro (Lake Skutari). It has an obvious raised metatarsal tubercle.

■ **Cretan Water Frog** *Rana cretensis* (range in blue) is similar and is endemic to Crete.

■ **Karpathos Water Frog** *Rana cerigensis* (range in orange) occurs on Karpathos and Rhodes and is the only water frog present on these islands.

■ **Levant Water Frog** *Rana bedriagae* (range in brown) inhabits Turkey and the Greek islands off the Turkish coast, and has the invalid synonym *Rana levantina*. The taxonomy of Turkish water frogs is not yet clear and *Rana caralitana* may constitute another valid species.

Levant Water Frog from Naxos

Iberian Water Frog
Rana perezi
Typical frogs RANIDAE

> ♂ 4–7 cm, ♀ 5–10 cm
> Water frogs
> Metatarsal tubercles small and low; backs of thighs
>   not yellow; Iberia and southern France

Description Mid-sized water frog with small, low metatarsal tubercles. Very similar to Marsh Frog (see p. 106), but mostly smaller, more slender and with more pointed snout. Upperside smooth or slightly granular, rarely warty and with two obvious, mostly bronze, dorsolateral folds. Coloration of back variable: green, brown, grey or yellow with irregular dark spots and often a yellow central dorsal line. Belly whitish with pale grey patches. Well-developed webbing. Back of thighs whitish with brown barring and with no yellow. Males have grey lateral vocal sacs and dark nuptial pads during mating season.

Distribution and habitat Found in Iberia and southern France (east to Lyon) in most habitats except for high mountains. Has also been introduced to the Canary Islands, Azores and Madeira, and possibly also to south-east England and the Balearic Islands. Found especially at altitudes up to

1,800 m and rarely up to 2,400 m. Very adaptable species which occurs in still wetlands of various sizes, fast-flowing streams that are not too cold and even in brackish lagoons close to the coast.

Notes Nocturnal and diurnal, often rests on banks, from where it jumps into the water when a predator approaches. Breeds from February–November depending on region and altitude. 1,000–10,000 eggs are deposited in small clumps of spawn on sub-aquatic vegetation.

Subspecies In the Sierra de Gredos of central Spain populations of the Iberian Water Frog have a very warty, almost toad-like, skin texture and can reach an amazing size of about 14 cm. The taxonomic status of this

Head of a typically coloured Iberian Water Frog

population is requires further study and clarification.

Additional species

■ **Graf's Hybrid Frog** *Rana* kl. *grafi* (range in blue) is a hybrid form (see p. 105 for details of the classification of hybrids) with Marsh Frog as one of the parents. It is almost indistinguishable from *R. perezi* and has only rarely been identified with certainty in southern France and Spain. It may be more widely distributed.

Pair of Iberian Water Frogs in amplexus (Sierra de Gredos)

Italian Pool Frog
Rana bergeri
Typical frogs RANIDAE

> ♂ 4–7 cm. ♀ 5–8 cm
> Water frogs
> Large semicircular metatarsal tubercle; back of thighs strongly yellow; only Italy and Corsica

Description Small, green frog, similar to Pool Frog (see p. 102), but with longer hind legs. Upperside green to brownish, with round, black spots and often a pale green central dorsal line. Underside whitish, with some grey patches. Backs of thighs strongly yellow to orange and barred brown. Males have white lateral vocal sacs. Metatarsal tubercles are large and semicircular.

Distribution and habitat In Italy south of the Po Estuary, introduced to Corsica. In open lowlands and hilly habitats on small, well-vegetated stagnant water bodies, ponds and ditches.

Notes See water frogs (p. 102–111).

Subspecies Individuals in the Abruzzi reach 11 cm and have a dark belly. Their status is unclear.

Additional species

■ **Italian Hybrid Frog** *Rana* kl. *hispanica* (range hatched blue) is a hybrid (see p. 105) between *R. bergeri* and Marsh Frog *R. ridibunda* (pure Marsh Frog populations do not occur in Italy).

American Bullfrog
Rana catesbeiana
Typical frogs RANIDAE

> 10–20 cm
> Bullfrogs
> Very large; no dorsolateral folds; large eardrums
> which are one to two times the diameter of the eye

Description Extremely large frog; may weigh up to 0.5 kg. Olive-green to brown above with black spots and no pale central dorsal line or dorsolateral folds. Underside white with grey patches. Males have lower vocal sac. Large eardrums twice as wide as the the diameter of the eye in males and at least as wide in females. Has strong webbing.

Distribution and habitat Native to eastern North America (west to the Rocky Mountains), occurs worldwide today, for example Brazil and Europe (upper Rhine Valley, Italy, France). Found in large, quiet and slow-flowing bodies of water with sunny areas. Hibernates in the water.

Notes Diurnal and nocturnal, strongly associated with water. Breeding in May and June, lays up to 20,000 eggs. Competes strongly with European species. Huge tadpoles grow to 10-17 cm before metamorphosing into 6–7 cm long frogs

Similar species European water frogs have smaller eardrums.

Hermann's Tortoise
Testudo hermanni
Tortoises TESTUDINIDAE

> ♂ 20–31 cm, ♀ 20–36 cm
> Tortoises
> Horny scale on tail-tip; no spurs on thighs;
> often two supracaudal plates

Description The shell length of adults is normally under 25 cm (♂) or 30 cm (♀). Domed carapace with oval outline that may appear lumpy. Toes have strong, separate claws, which are grown together and form a 'club-foot'. Head and limbs olive yellow. Carapace yellowish, brownish or orange-greenish yellow, with dark brown to black spots, especially on the front and the sides of the vertebrals and costals (back fringe of the marginals is often pale). Often has dark diagonal markings on the marginals. Plastron yellow with dark patches, which form two parallel and nearly continuous longitudinal bands in the western nominate subspecies. In the eastern subspecies (*T. h. boettgeri*) these are irregular or absent. Appearance is similar to the Spur-thighed Tortoise (see p. 116), but has a large scale on tail-tip and no spurs on the thighs. Usually there are two supracaudal plates, but this is not always the case and is therefore not a reliable single feature for certain identification. Rear marginal scales never flared out as in other species. Males with concave plastron and longer tail that is broadened at the base.

Distribution and habitat Widely distributed in the Balkans and Italy, including some Mediterranean islands. Isolated populations in north-east Spain and southern France. Sunny, warm, but not too dry habitats up to 1,500 m, common in open deciduous forests, scrub, meadows with shady places or sunny, well vegetated hills.

Notes Diurnal, likes the sun. Hibernates underground from October– March. Males have fierce fights ramming their bodies during spring and sometimes autumn mating. This can be heard from a great distance. Mating noisy with strange beeping sounds. Eggs are laid May–July on sunny, exposed hills. Clutches of about 5–10 eggs. In captivity can reach 50 years of age.

Subspecies *T. h. hermanni* in western part of the range (Spain, southern France and Italy, including the large islands) reaches length of 20–25 cm and shows a contrasting pattern of yellow and black as well as yellow cheeks. In contrast, *T. h. boettgeri*, from the south-eastern Mediterranean, is

Juvenile Hermann's Tortoise

darker, brownish or olive and might reach an exceptional length of 30 cm. The taxonomy is still under discussion. Some authors define an additional subspecies *T. h. hercegovinensis* from the north-eastern Adriatic coast; some treat it as a separate species.

Similar species Spur-thighed Tortoise (see p. 116) and Marginated Tortoise (see p.118), lack the horny scale on tail-tip.

Rear view showing characteristic horny scale or spike on the tail and divided supracaudal plate

Spur-thighed Tortoise
Testudo graeca
Tortoises TESTUDINIDAE

> ♂ 15–25 cm, ♀ 20–36 cm
> Tortoises
> No horny scale on tail-tip; spurs on thighs;
  supracaudal plate usually undivided

Description Average length of the domed, oval carapace usually lies well below the maximum lengths given. Similar to Hermann's Tortoise (see p. 114), with which it can be confused. Central vertebral plates broader and larger, scaling of front limbs coarser, mostly in form of four rows of scales on the arms. There is no horny scale on the tail-tip, and there is an obvious spur on each thigh. The supracaudal plate is usually undivided, but sometimes there may be a central dividing line. Carapace is variable yellow, brownish olive or buff. Western nominate subspecies quite bright with no or just a little black inside the plates and the lines bordering the plates of the carapace. The eastern subspecies *T. g. ibera* shows intense pigmentation, especially at the front. Plastron with diffuse dark patches or uniform yellow. Males with concave plastron and longer, strong tail. Rear of the carapace flares out in older individuals.

Distribution and habitat Two subspecies: *T. g. graeca* at scattered locations along the western Mediterranean coast up to 800 m; *T. g. ibera* on the south-eastern Balkan peninsula at altitudes of up to 1,300 m. Introduced to different locations in Europe. In dry, semi-open or barely forested habitats with sunny and shady places, also fields and gardens.

Notes Mating takes place in spring with noisy, high-pitched sounds during courtship (see Hermann's Tortoise, p. 114, and Marginated Tortoise, p. 118). One or two clutches of 4–8 eggs are laid in holes in the soil in sunny places.

Subspecies The western form of the two European subspecies, *T. g. graeca*, usually has a bright yellow to buff carapace that is not longer than 20 cm and which has a contrasting pattern. It occurs in small populations in southern Spain and on the Balearic Islands. The eastern subspecies *T. graeca ibera* is

With age, the marginal plates often form a concave edge

distributed from Greece, Bulgaria, Romania and eastwards to the Caucasus (the name of the subspecies *ibera* refers to Caucasian Iberia and not the Iberian Peninsula). It is generally darker, and on average larger, reaching 20–30 cm. The two subspecies have been treated as separate species in the past; additional species are described from Africa and western Asia.

Similar species Hermann's Tortoise has a horny tip on its tail, Marginated Tortoise lacks spurs on the thighs.

Note undivided supracaudal plate and tail without horny tip

Marginated Tortoise
Testudo marginata
Tortoises TESTUDINIDAE

> 20–40 cm
> Tortoises
> No horny scale on tail-tip; no spurs on thighs; supra-
  caudal plate undivided; rear marginals bent upwards

Description Long and domed carapace, which is flared at the ends and appears to have a waist in the middle. Old individuals in particular have enlarged outer marginals, which stick out sideways like the rim of a hat. Rear of carapace has an obvious ragged fringe, except for animals from Sardinia. No horny scale on tail-tip, usually with undivided supracaudal plate and without obvious spurs on the thighs (small inconspicuous scales may occur). Carapace relatively dark, mostly dark brown to black except for the rear parts of the marginal plates and the yellowish to orange centres of the central and vertbral plates. Pale markings on the scales become darker with age, so that old individuals may be completely dark. Plastron ochre to yellowish with a dark pattern resembling a pennant. Males with longer and stronger tail, concave plastron and marginal plates bent prominently upwards.

Distribution and habitat Small distribution in the south of Greece (south of the Olympus mountains), especially on the Peloponnese and some neighbouring islands (e.g. Skyros). Introduced on Sardinia a long time ago (especially in the north east), also introduced to some parts of Italy (west coast, east of Po delta). In hilly and mountainous regions up to 1,600 m, especially in deciduous forests, olive groves, stony hillsides or in open landscape with small trees and bushes.

Notes Like all European tortoises active during the day and feeds mostly on plants. Sometimes takes small invertebrates such as slugs and worms, and occasionally even excrement is eaten. Movements slow and deliberate. When in danger the animal draws back into the shell and closes the opening with its strong front limbs. See also Hermann's Tortoise and Spur-thighed Tortoise (p. 114–117).

Similar species Hermann's Tortoise with horny scale on tail-tip, Spur-thighed Tortoise with obvious spurs on thighs.

Additional species

■ **Weissinger's Tortoise**
T. weissingeri (range in blue), occurs only in the southern Peloponnese. It's status is still debated by taxonomists: it is possibly a subspecies *T. marginata weissingeri*. Markings

Adult Marginated Tortoise in its natural habitat

on the carapace are often blurred and indistinct. Eggs are long and oval, not conical and the animals are a bit smaller with a length of 20–25 cm.

Weissinger's Tortoise

Spanish Terrapin
Mauremys leprosa
Old World terrapins BATAGURIDAE

> 13–23 cm
> Terrapins
> Yellow stripes on head and neck of equal width;
> upper jaw edge smooth; eye pale

Description Similar to Balkan Terrapin (see p. 122), but generally a bit heavier and paler, with stout head, smooth (not ragged) edges of upper jaw and pale eye. Pale yellow to reddish streaks are of equal width on neck and head. In contrast to the Balkan Terrapin, the stripes on the next have slightly paler centres and there is often a pale spot between the eardrum and eye. Limbs dark olive with yellow or orange streaking. Carapace and plastron connected by a bony bridge. Carapace low and long and typically has several domed plates. The carapace of young animals has an obvious central keel and is weakly developed, with mostly disconnected keels on sides. Young individuals often have pale orange centres to the costal and marginal plates. In old individuals, the upper shell is divided by a central keel at least at the rear. The basic coloration of the upperparts is pale brown to olive-brown. Plastron yellow to pale brown; often with black patches in small individuals and all black in old individuals.

Distribution and habitat On the Iberian Peninsula as well as three areas in southern France (western Pyrenees, Banyuls, Hérault); also in north-west Africa. Most common in lowland areas in large, warm and slow flowing bodies of water rich in vegetation, such as lakes or quiet parts of rivers and streams, but also in the smallest temporary puddles and even polluted or brackish places. In Europe found up to 1,100 m, in Africa up to almost 2,000 m. Usually has a short period of hibernation in the mud on the bottom of lakes.

Notes Diurnal and very timid species that likes to rest on sun-exposed banks of ponds. When in danger disappears into the water and hides. If caught by a predator, glands emit a musk-like substance. Depending on availability, these animals feed on animal and plant matter, e.g. aquatic plants, algae, water insects, snails, tadpoles or dead fish. Mating occurs from the beginning of March, and eggs are laid between April and August, often with two laying periods separated by a break of about one month. The clutch size is 4–13 eggs; young animals hatch after three months.

Similar species European Pond Terrapin has many spots and patches but lacks streaks on neck. Balkan Terrapin has stripes on the neck that are thinner than the upper central stripe.

Young animal with obvious keels

Balkan Terrapin
Mauremys rivulata
Old World terrapins BATAGURIDAE

> 12–21 cm
> Terrapins
> Pale stripes on neck broader than on head;
  upper jaw ragged; eye dark

Description Similar to Spanish Terrapin (see p. 120), but more slender and with smaller head and dark eye. Upper jaw has fine ragged edges. Yellowish streaking of the neck reaching the cheeks (reaching the snout in the closely related Caspian Terrapin, *M. caspia* from Asia), and the outer streaks are always narrower compared to the central, upper streaks. Limbs are dark olive with a relatively obvious striped pattern. Carapace and plastron

rigidly connected by bony bridge. Low or slightly domed, long carapace with relatively uniform, pale, dark or olive-brown basic coloration, which varies depending on the conditions of the surroundings (colour can be affected by minerals containing sulphur or iron). Young animals very dark with irregular, bright pattern of interconnected lines that gradually disappears in older individuals. Marginal plates on the upper side have indistinct, brownish eye-shaped pattern. The bony bridge between carapace and plastron appears completely black. Plastron black except for paler seams.

Distribution and habitat Found in warm areas of the southern Balkans (including many Aegean islands) and the neighbouring Middle East as far as Israel, usually not higher than 500 m. Occurs especially in slow-flowing and quiet bodies of water rich in vegetation close to the coast, such as streams, rivers, freshwater lagoons or ponds with muddy substrate on the bottom. Also in polluted bodies of water and brackish estuaries.

Notes Previously regarded a subspecies of Caspian Terrapin, *M. caspica*, which has a more easterly distribution. Diurnal, favouring sunny places and hiding underground when the water has evaporated. Very timid, the animal disappears into the water and hides under a layer of mud. If caught by a

Adult Balkan Terrapin sunbathing

predator, glands emit a musky substance. Diet is seasonal and consists mainly of water plants, algae, leaves and fruit; also takes small water insects, slugs and worms. Mating takes place in spring, egg laying is usually in June and July; may lay eggs twice a year, especially in the south of its range. Clutch size usually 4–10 eggs, and the offspring hatch after another 3–4 months.

Similar species European Pond Terrapin lacks streaking on the neck, Spanish Terrapin has dark and pale streaks of equal width on neck.

Newly hatched juvenile

European Pond Terrapin
Emys orbicularis
Pond terrapins EMYDIDAE

> 10–23 cm
> Terrapins
> Neck has yellow spots and lacks streaking; carapace and plastron only loosely connected; eye pale

Description Small to mid-sized terrapin of varying length; in the Mediterranean area mostly 12–16 cm long, in central and eastern Europe on average a little larger. Some populations have tiny individuals, just 10–12 cm long, e.g. on the Peloponnese or along the Dalmatian coast. Head and neck dark brown with a variable pattern of pale spots and patches, which is better developed in females and may be completely absent in old males; there are never longitudinal stripes as in members of the genus *Mauremys*. Shell oval and slightly domed. In contrast to the genus *Mauremys*, carapace and plastron are connected by movable bony cartilages. Old individuals have an additional diagonal hinge on the plastron, which allows the front of the body to be movable. Carapace dark brown to black and, depending on the subspecies, with a variable pattern of whitish to yellow dots and streaks, often arranged like a star. Pale coloration disappears with age and fully grown individuals appear almost black. Plastron yellow and with variable pattern of black patches, which depends on the age and is characteristic for each sub-species. *E. o. orbicularis* has a more or less uniform black plastron, *E. o. fritzjuergenobsti* has an almost completely yellow plastron. Toes with strong webbed claws. Tail fairly long. Males are generally smaller than females, with a concave plastron and strong base to the tail.

Distribution and habitat Many subspecies are distributed across north, east and south Europe, and also in north-west Africa and parts of west Asia eastwards to the Aral Sea. The European Pond Terrapin has one

European Pond Terrapin of the subspecies hellenica *from Greece*

of the largest distributions of the 300 recognized species of turtles and terrapins and is the only European representative of a family that is otherwise only found in the New World. In Spain these animals are found up to an altitude of 1,000 m, but most populations live below

Individual with typical pattern of yellow streaks and spots

Head of European Pond Terrapin showing yellow spotting on neck and pale eye

400 m. Many northern European populations can be traced back to introductions from the Mediterranean area or eastern Europe (e.g. Hungary). The species is found in large, well-vegetated, quiet waters, such as ponds, lakes or flooded pastures along rivers, which have suitable places for sun-bathing and egg-laying. Also occurs in irrigation systems, and occasionally (especially in southern Europe) in rivers and streams, sometimes in brackish water. Hibernates in mud at the bottom of ponds.

Notes Very shy and timid diurnal species that likes the sun. If caught, glands emit a musky smelling substance. In northern and central Europe active from March–November. Diet: insects which fall into the water, small crustaceans, water insects and larvae, dead fish and other carcasses, and amphibians and their larvae. Juveniles also eat plants. Mating takes place from March–May. Mating takes about two minutes and the male wraps his limbs around the female's shell. The female lays one or two clutches of hard-shelled eggs between May and June. She digs holes, often far away from the water, and deposits 5–20 eggs in them. The young hatch after about two months, depending on the temperature. The sex is mainly determined by the incubation temperature. At 28°C, all offspring will be male; at 30°C females hatch. They mature in 5–10 years and can live to 70 years.

Subspecies Presently 13 subspecies are recoginzed. These are divided into six groups, the distribution and systematic classification of which is not completely clear. The 'orbicularis group' consists of *E. o. orbicularis* (from central and eastern Europe), *E. o. colchica* (from the south-east Balkans) and *E. o. eiselti* (from Anatolia); the 'hellenica group' of *E. o. hellenica* (from the Balkans and Crimea); the 'iberica group' of *E. o. iberica* and *E. o. persica* (both from western Asia); the 'occidentalis group' *E. o. occidentalis* (north Africa), *E. o. fritzjuergenobsti* and *E. o. hispanica* (both from

Spain); the '*galloitalica* group' of the very closely related and still debated subspecies *E. o. galloitalica* (from southern France and Italy), *E. o. lanzai* (from Corsica) and *E. o. capolongoi* (from Sardinia); and the '*luteofusca* group' of *E. o. luteofusca* (central Anatolia).

Similar species Spanish Terrapin and Balkan Terrapin both have obvious dark and pale streaking on the neck.

Additional species

■ **Red-eared Terrapin**
Trachemys scripta can grow up to 30 cm (female), and is easily identified by the red patch on the cheeks. It is indigenous to North America but has been introduced to many European

Red-eared Terrapin with typical red patches

countries and has thriving populations in parts of Britain, Spain and Germany. It is found in large ponds, lakes and slow-flowing rivers.

European Pond Terrapins from the Dalamation coast can be quite small

Loggerhead Turtle
Caretta caretta
Sea turtles CHELONIIDAE

> 80–130 cm
> Sea turtles
> Carapace with five pairs of costal plates; lives in the sea; paddle-like forelimbs with two claws

Description Large sea turtle with average shell length 80–100 cm (weight 100–150 kg). Oval, red-brown upper shell with five pairs of costal plates, often with attached crustaceans. Plastron yellowish. Juveniles have three keels on carapace. Low, paddle-like forelimbs with two claws.
Distribution and habitat Found worldwide. Lays egg on sandy beaches in tropics, sub-tropics and temperate zones, including the eastern Mediterranean.
Notes Rare and very endangered.

Diet: jellyfish, crustaceans and other invertebrates.
Additional species
■ **Kemp's Ridley Turtle** *Lepidochelys kempii* is similar but shell grey-brown. Rare visitor to Atlantic coast. Breeds Mexico.
■ **Green Turtle** *Chelonia mydas* has four pairs of costal plates. Breeds Cyprus, rare elsewhere in Mediterranean and Atlantic.
■ **Leathery Turtle** *Dermochelys coriacea* (Dermochelyidae). Very rare visitor, 150–200 cm long, leathery skin covering carapace.

Starred Agama
Laudakia stellio
Agamas AGAMIDAE

> 20–35 cm
> Agamas
> Head with small scales rather than large plates; spiny scaling on body; tail conspicuously barred

Description Mid-sized, robust; body 10–15 cm, tail variable (often a short stump). Small scales on head, no large plates. Spiny scales (especially on nape and limbs) and tubercles form rows on back and flanks. Tail barred with spiny scales. Upperside dark grey, brown or yellow-brown with faint yellow diamond markings or rows on centre of back. Dominant males bright blue, orange-red or yellow; often bob their heads. Belly pale brown and barred.

Distribution and habitat Found in northern Greece (around Thessaloniki) and on some Aegean islands (e.g. Naxos, Mykonos, Lesvos, Rhodes). Introduced to Malta and Corfu. Favours dry, open habitats such as ruins, rocks and walls.

Notes Diurnal. Likes sunny places and can often be seen running around on the ground. When faced with danger may climb trees or escape into crevices. Lays about 10 eggs which hatch after 2–4 months.

129

Mediterranean Chameleon
Chamaeleo chamaeleon
Chameleons CHAMAELEONIDAE

> 20–30 cm
> Chameleons
> Body laterally flattened; bulging eyes independently movable with ring-shaped eyelids; prehensile tail

Description Unmistakable due to laterally flattened body with low back crest and conspicuous helmet on head, as well as a 10–12 cm long prehensile tail. Fingers and toes grown together to form prehensile 'pliers'. Very bulging eyes which can move independently and which are protected by ring-like scaly eyelids. Variable coloration and patterning of back, mostly camouflaged olive to pale green, brown or grey, with two whitish stripes on the sides and irregularly formed dark patches. Can undergo extreme, rapid changes of coloration in response to different moods. If excited, the animal often appears brighter, whereas they are usually paler at night. **Distribution and habitat** Large range stretches from Morocco along the south Mediterranean coast to the Middle East and on to western India. Within Europe just a few isolated populations are found in the south of the Iberian Peninsula (in the

Portuguese province of Faro and the Spanish provinces of Huelva, Cádiz, Málaga, Granada and Almería) as well as on Malta, Crete, Chios and Samos (but *not*, as often stated, on Sicily and in the Peloponnese). Most populations in Spain, Portugal, Malta and Crete are probably due to introductions during ancient times. In Europe found in well-vegetated areas close to the coast, in the Spanish province of Málaga up to an altitude of 900 m, and outside Europe up to 2,600 m. Mostly on bushes and in trees, in open woods with acacia, tamarisk or broom. In parts of northern Africa lives on the ground. In Iberia it may occur in pine forests or close to settlements, for example in plantations, olive groves or eucalyptus woods. **Notes** Climbs bushes to hunt prey. It catches insects and spiders by shooting out its tongue, which is about the length of the animal itself, at great speed. It focuses on its prey, aims accurately and shoots out its tongue from its half-opened mouth. The sticky tip of the tongue connects with the prey. Mating takes place in August and September, the female digs a hole and deposits 20–30 eggs. The young hatch after 8–9 months.

Additional species

■ **African Chameleon**
Chamaeleo africanus (range in

Young Mediterranean Chameleons are 6cm long on hatching

blue) occurs in mainland Greece south of the Peloponnese. It can be distinguished from the Mediterranean Chameleon by its bigger size (up to 46 cm) and a small spur on the hind leg. This population may have been introduced from Egypt in ancient times.

African Chameleon from the Peloponnese

Moorish Gecko
Tarentola mauritanica
Geckos GEKKONIDAE

> 12–15 cm
> Geckos
> Pupils vertical and slit-like; toes have undivided adhesive pads; obvious claws on third and fourth toes

Description Largest European gecko with large, flattened head and robust body. Tail almost as long as body. Large, protruding eyes with vertical, slit-like pupils and eyelids grown together to form transparent 'glasses'. Back with rows of large, keeled tubercles creating a spiny appearance; tail flattened with spiny tips to the outermost scales. Upperside grey or brownish with dark, faded barring, especially well developed on tail and in juveniles.

Regenerated tails are uniform and have no spiny tubercles. Belly yellowish. On the undersides of each toe are 12 large, oval, undivided adhesive pads. Claws are only visible on the third and fourth toes. The species' amazing ability to 'stick' to vertical structures is due to the molecular interactions between the surface upon which it is standing and tiny hair-like structures on the underside of the pads. The toes of dead geckos still stick to walls.

Distribution and habitat Occurs in coastal areas around the Mediterranean, including northern Africa and most islands. Also in central areas of the Iberian Peninsula and Italy. Introduced to many other places worldwide, e.g. Montevideo (Uruguay) and California (USA). Prefers dry, warm, stony areas of lowland and hilly regions, in the Spanish Sierra Nevada up to 2,350 m. Closely associated with human settlements and can often be found on walls, under roofs and in piles of wood or rocks and caves or crevices.
Notes Climbs very well. Adhesive pads enable it to adhere to smooth surfaces, overhanging walls and panes of glass. Mostly nocturnal, but may be diurnal and basks in the sun for hours. Diet: mainly insects and other arthropods, occasionally small lizards caught close to well-lit places. Utters squeaking noises if in distress and can shed its

Underside of front foot showing adhesive pads and claws on third and fourth toes

tail (autotomy). During the breeding season may live in pairs for a short period. In the summer, two eggs are deposited in a crevices or a hole in the ground. Juveniles hatch after 5–12 weeks, depending on the temperature.
Subspecies The Moorish Gecko is divided into four subspecies and only the nominate *T. m. mauritanica* inhabits Europe.
Similar species Other geckos, see Turkish Gecko (p. 134).

Moorish Gecko from Menorca

Turkish Gecko
Hemidactylus turcicus
Geckos GEKKONIDAE

> 8–11 cm
> Geckos
> Pupils slit-like, vertical; adhesive pads do not reach
tips of toes, divided on lower surface; toes with claws

Description Mid-sized, slender gecko with relatively long tail, about 60 per cent of the complete body length. Eyes with slit-like, vertical pupils and eyelids joined together to form transparent 'glasses'. Back with 14–16 rows of large tubercles separated by many small scales. Several rows of spiky scales on tail. Upperside only weakly pigmented, appearing transparent, flesh-coloured, yellowish, reddish or brownish with irregular dark spots. The tail has dark barring and a very contrasting pattern is especially clear on immatures. Underside is whitish. Flat, well-developed adhesive pads on the underside of the toes and fingers, which are divided on the lower surface and do not reach the tips of the toes (small claws visible on all toes). Extraordinary ability to adhere to and climb on vertical, smooth and overhanging structures such as walls, rocks, ceilings and glass windows, as described under Moorish Gecko.

Distribution and habitat Along the Mediterranean coast of Europe and Africa, as well as most of the Mediterranean islands; also introduced to the Canary Islands and Central and North America. Found in dry and warm areas with many hiding places, especially in coastal habitats, in Spain up to 1,200 m. On boulders, old trees, piles of wood or dry-stone walls. Frequently found close to or in human settlements.

Notes Agile gecko which can climb very well and which can give audible, cat-like sounds. Usually nocturnal but can often be seen sunbathing. Diet: small spiders and insects that are often caught under artificial lighting at night. Mating takes place in early summer and two eggs are deposited under stones or in the earth.

Divided adhesive pads on the undersides of the toes

Subspecies Two subspecies recognized, with the nominate *H. t. turcicus* occurring in Europe.

Similar species Moorish Gecko stronger with large, undivided adhesive pads that reach the toe tips. European Leaf-toed Gecko with leaf-like pads at toe tips. Kotschy's Gecko with strongly kinked toes.

Turkish Gecko from the Greek island of Naxos

Kotschy's Gecko
Cyrtopodion kotschyi
Geckos GEKKONIDAE

> 9–13 cm
> Geckos
> Pupils vertical, slit-like; toes kinked upwards and lack adhesive pads; back and tail have spiked tubercles

Description Delicate with quite long tail (60 per cent of the complete length) with spiky tubercles (unless regenerated). Vertical, slit-like pupils. Back with longitudinal rows of tubercles. Upperside yellowish, brownish or dark grey with black barring (especially on tail, particularly in juveniles). Belly whitish to brownish. Toes kinked upwards at tips and have no adhesive pads.
Distribution and habitat South-east Italy and south-eastern Balkans, especially at lower altitudes, in south up to 1,400 m. On boulders, dry-stone walls, houses and old trees.
Notes Some authors place this species in the subgenus *Mediodactylus*. Agile, at home in tree or on the ground. Diurnal and nocturnal; likes sun-bathing. Diet: small arthropods. Lays two eggs under stones.
Subspecies Over 20 subspecies described from Aegean Islands, also from Greek mainland, e.g. *C. k. bibroni* from Peloponnese.
Similar species Turkish Gecko.

European Leaf-toed Gecko
Euleptes europaea
Geckos GEKKONIDAE

> 6–8 cm
> Leaf-toed geckos
> Pupils vertical, slit-like; pairs of heart-shaped adhesive pads on tips of toes; back and tail smooth

Description Smallest European lizard. Thick tail (especially if regenerated) not longer than 50 per cent of body length). Pupils vertical and slit-like. Back finely scaled, appearing smooth and without tubercles. Upperside brownish, yellowish or grey with diffuse patterning from pale and dark patches. Belly whitish. Flat, leaf- or heart-shaped pads at the tips of the toes

Distribution and habitat Has a small distribution on the Tuscan archipelago and adjacent larger islands (e.g. Sardinia and Corsica), and in some locations along the Italian coast up to 600 m. Numbers declining. Found under stones, loose bark, leaf litter, rarely on houses.

Notes Until recently classified under the genus *Phyllodactylus*. Nocturnal. Diet: small insects, spiders and isopods. Distress call a repeated squeak. Eggs, in clutches of 1–2, often laid in crevices by groups of females. Young 3 cm long on hatching. Can reach 20 years of age.

137

Sand Lizard
Lacerta agilis
Lacertid lizards LACERTIDAE

> 20–28 cm
> Green lizards
> Very variable; narrow scales on centre of back; flanks of ♂ mostly green with pale ocelli

Description Medium-sized, stout lizard with strong head and short legs. Body length 9–11 cm, tail about 1.5 times body length. Serrated collar. Centre of back has an area of narrow, longish, keeled scales that are clearly different from the larger dorsal scales. Keeled scales on tail. Back brown to grey-brown, often with three paler longitudinal stripes partly broken and interspersed by additional whitish spots or streaks; in between are several rows of black spots. On the flanks there are white ocelli that are fringed black or with dark barring. Some individuals in mainland Europe (but not in Britain) have a uniform reddish brown back (example shown in main picture) and sometimes almost uniform brown, green or black individuals may occur. Belly whitish or yellowish in female, mostly green in male

138

and often spotted dark. During courtship males have bright green flanks and throat, sometimes with a blue throat. All-green individuals may occur. Females usually brown but may have green flanks. Juveniles brown with whitish spots and three longitudinal stripes bordered by black spots.

Distribution and habitat The most widely distributed species of lizard apart from Common Lizard. Occurs from southern England to Lake Baikal, from southern Sweden to northern Greece. In England rare and restricted to dunes and heaths. Elsewhere can be common in lowland and hilly country, but declining in parts of range. Does not occur at very high altitudes (up to 2,200 m in the Spanish Pyrenees). Habitat varied, but especially in semi-open county such as meadows, heathland, forest edges or railway embankments, also in gardens, vineyards and quarries. In the south inhabits mountain ranges, open slopes and rocky places.

Notes Diurnal and sun-loving. Mating takes place in spring; often each female mates with several males. As with some other species of lizards, a remarkable process of mate selection takes place, where the most genetically different males are chosen by the females (the mechanism for this is unknown). As mating with genetically

Female showing ocelli on flanks

similar males leads to fewer offspring than would statistically be expected, this ensures that interbreeding is less likely to happen. Between May and July about 5–15 eggs are laid in the ground once or twice a year; the juveniles hatch after 7–10 weeks.

Subspecies The nominate subspecies *L. a. agilis* occurs in western and central Europe. Elsewhere, eight additional subspecies are recognized, three of which occur in the area covered by this book: *L. a. argus* in eastern central Europe, *L. a. bosnica* in the south-west Balkans and *L. a. chersonensis* in eastern Poland, Romania and northern Bulgaria. The Spanish subspecies *L. a. garzoni* is no longer considered valid.

Similar species See other green lizards, pp. 140–147.

Head of an adult male

Eastern Green Lizard
Lacerta viridis
Lacertid lizards LACERTIDAE

> 30–40 cm
> Green lizards
> Back bright green; throat blue in ♂; young juveniles brown on head; six longitudinal rows of belly scales

Description Large, elegant, long-legged lizard with pointed head. Body length 10–13 cm, tail about twice as long. Until recently lumped with the Western Green Lizard *L. bilineata* (and known simply as Green Lizard *L. viridis*). The two species cannot usually be reliably identified in the field, so the following description also applies to *L. bilineata*. Back of males and most females more or less uniform grass-green or yellowish green with very fine black spots, which are evenly spread. Some females and juveniles may have a brown back. Females can have two or four narrow whitish streaks along the flanks. Males, and some older females, have a bright blue throat and cheeks in breeding season. Belly plain whitish to yellowish green. Compared to the similar but strikingly larger Balkan Green Lizard, there are just six longitudinal rows of scales on the belly and the number of lateral scales on the head is lower (usually less than

140

20 temporal scales and no continuous row of supraciliary granules between supraoculars and supratemporals). Nostril is usually separate from rostral scale (no contact).

Distribution and habitat Balkans east to Ukraine and Turkey, but not on Aegean Islands. Isolated populations in Czech Republic and eastern Germany. Lowland species in north of range; occurs up to 2,000 m in south. Favours dry, warm, south-facing hills, sunny forest edges, meadows with lots of bushes, vineyards and rocky places.

Notes See *L. bilineata*, p. 142.

Subspecies Nominate subspecies in eastern Europe and most of the Balkans, *L. v. meridionalis* in north-east Greece and Black Sea coast. The recently described *L. v. guentherpetersi* occurs on the Aegean island of Evia (probably also on the adjacent mainland south of the Olympus mountains). Two additional subspecies live outside Europe.

Similar species Usually cannot be

Male and female of the subspecies L. v. meridionalis *sunbathing*

distinguished from *L. bilineata* in the field: only newly hatched juveniles of *L. bilineata* show an intense green coloration on head and neck (brownish or buff in *L. viridis*). Balkan Green Lizard has eight longitudinal rows of scales on belly, nostrils in contact with the rostral scale and the males often with a yellow throat. Schreiber's Green Lizard has black patches on the underside. Sand Lizard has central scales on back that are distinctly smaller than those on the flanks.

Male of the nominate subspecies from the Czech Repubic

Western Green Lizard
Lacerta bilineata
Lacertid lizards LACERTIDAE

> 30–40 cm
> Green lizards
> Back bright green; throat blue in ♂; head of juvenile greenish; six longitudinal rows of scales on underside

Description The Western Green Lizard was previously lumped with the Eastern Green Lizard as a single species (known as Green Lizard), but its status as a separate species has now been unanimously accepted. The body length is about 13 cm and the tail is about twice length of body. For a full description see Eastern Green Lizard, p.140.

Distribution and habitat Found from northern Spain across France to Italy and Slovenia, with a few isolated populations in western Germany. Does not occur on most of the western Mediterranean islands (except for Elba). Found especially in areas with a warm climate at lower and mid altitudes; in the Italian Abruzzo region and northern Spain it occurs up to 2,160 m. Further north it inhabits dry and warm areas with southward-facing hillsides, especially vineyards, dry meadows, areas of boulders, places overgrown with bramble bushes and sunny forest edges.

In the south it often favours ravines in moist areas of mountain ranges.

Notes Diurnal, timid, very agile and fond of the sun. When in danger scurries noisily into a hole or crevice. Can shed tail (autotomy). Diet: mainly insects, spiders, slugs, worms and occasionally fruit or young reptiles or birds. During mating season in April and May there are often fights between males. About 1–2 months after mating the clutches of 5–20 eggs are deposited. If conditions are good a female may lay twice a year. The young hatch between August and October.

Subspecies The nominate subspecies is found across much of western Europe (from northern Spain across France to Italy, and in western Germany), while in *L. b. chloronota* from southern Italy (Sicily and

Juvenile Western Green Lizard showing diagnostic green head

Calabria) the breeding male has blue sides to the head. Two additional subspecies are still considered valid: *L. b. fejervaryi* from central Italy and *L. b. chlorosecunda* from the Italian Apulian region.

Similar species Other species of green lizards. See Eastern Green Lizard (p. 140).

Typically coloured female of the Italian subspecies fejervaryi

Balkan Green Lizard
Lacerta trilineata
Lacertid lizards LACERTIDAE

> 40–60 cm
> Green lizards
> Back pale green; throat yellow in ♂ ; juvenile has 3–5 stripes; eight longitudinal rows of scales on belly

Description Large, robust lizard with relatively long legs and long tail. Largest representative of its genus, body length 16–17 cm, tail 1.5–2 times body length. Similar appearance to Eastern Green Lizard and species may occur side by side, but the Balkan Green Lizard generally prefers warmer places. Back of adults more or less uniform, usually bright pale green or yellowish to brownish green with regular, fine black spots. Juveniles, sub-adults and some females brown with three or five narrow whitish longitudinal stripes down centre of back (often partly very faded) and along the flanks. Male has yellow or green throat, rarely blue on the sides of the neck during breeding season. Underside plain greenish to yellowish. Compared to Eastern Green Lizard there are eight longitudinal rows of belly scales on the underside. In most cases there are more than 20 temporal scales as well as a continuous row of eight supraciliary granules between the supraoculars and supratemporals. The nostrils usually reach the rostral scale.

Distribution and habitat Found in the south, west and east of the Balkans, especially in lower and middle altitudes up to 1,000 m. Also occurs in western Asia up to 2,000 m. Favours semi-open, dry and sunny areas such as olive groves, sand dunes, boulders or dry-stone walls and piles of wood, rarely in wet areas or close to streams.

Notes Very timid, diurnal species that hides as soon as danger is perceived. Climbs well (even on trees) and likes sunbathing close to thorny bushes that provide shelter when danger threatens. Diet: larger insects, spiders and slugs, also smaller lizards and young rodents. Mating takes place in spring after a hibernation period of 4–6 months. Clutches of 5–20 eggs are deposited (often twice a year) from May to June. The young hatch in August and September.

Subspecies The nominate subspecies *L. t. trilineata* occurs in the central and eastern parts of Greece, *L. t. major* in the western Balkans and *L. t. dobrogica* along the Romanian and Bulgarian coast of the Black Sea. Additional subspecies are distributed on the Aegean Islands: *L. t. citrovittata* on Tinos and Andros, *L. t. hansschweizeri* on the nearby islands of Milos, Kimolos, Serifos and Sifnos, *T. t. polylepidota* on

Balkan Green Lizard of the sub-species dobrogica *from Bulgaria*

Crete and Kythira and *L. t. diplochondrodes* on Rhodes, Kos and other islands near the Turkish coast. Additional subspecies are described from Asia.

Similar species See Eastern Green Lizard, p. 140.

Balkan Green Lizard of nominate subspecies from central Greece

Schreiber's Green Lizard
Lacerta schreiberi
Lacertid lizards LACERTIDAE

> 25–40 cm
> Green lizards
> Back of ♂ green, belly plain; juveniles have 3–5 rows of ocelli; eight longitudinal rows of scales on belly

Description Relatively large, stout green lizard with strong legs, broad head and long tail. Body length not longer than 13.5 cm, tail 1.5–2 times body length. Coloration of back variable in adults, males usually grass-green or yellowish green with numerous small and large blackish spots. Female mostly brownish or grey brown with green tones and irregular black spots, which are often close together and form three longitudinal rows; whitish ocelli may occur. Juveniles and sub-adults are also brown and show yellowish, black-bordered ocelli on the flanks forming three longitudinal rows; there are no longitudinal stripes. Males (sometime even older females) may have bright blue throat during breeding season. Belly yellowish and usually spotted black (especially in males). Eight longitudinal rows of scales along the underside. **Distribution and habitat** Only in the Iberian Peninsula (northern and central Spain, Portugal), in scattered isolated populations. Favours moist areas of hilly and mountainous regions up to 1,800 m. Prefers habitats rich in vegetation, e.g. overgrown verges, bramble thickets along paths and boulders among dense vegetation; often found close to mountain streams.

Notes Lethargic lizard that enjoys sunbathing and often dives for cover to escape from predators. Diet: larger arthropods, especially insects and spiders, but also slugs and smaller lizards. Mating in March after a long period of hibernation. The males occupy territories and ritualized fights may occur in which the blue throat is used as a means of intimidation. Clutches of 10–20 eggs are laid in May and June. Maturity is reached in 2–3 years and they can live for 10 years.

Similar species On the Iberian

Head of a breeding male

Peninsula only the Western Green Lizard, which has six pairs of longitudinal rows of scales on the belly and no dark patches underneath. Ocellated Lizard is much larger with ten pairs of longitudinal rows of scales on the belly and blue ocelli. See also Eastern Green Lizard, p. 140.

Brownish coloured female

Ocellated Lizard
Timon lepidus
Lacertid lizards LACERTIDAE

> 60–75 cm
> Ocellated lizards
> Back green, blue ocelli on flanks; very large, strong;
> 10 longitudinal rows of ventral scales

Description Largest and heaviest European lizard with a maximum body length of 20–24 cm, females usually smaller than 20 cm. The long tail, broad at the base, is 1.5– to 2 times body length. Stout, strong build and wide head, which seems to be enlarged around the cheeks. Coloration of the back variable, usually bright green, rarely yellowish or brown-grey (especially for regenerated tails), and sometimes with numerous dark spots forming patterns resembling rosettes. On flanks 3–4 rows of obvious deep blue patches bordered by black spots. Juveniles green to brownish on back, with conspicuous white, black-bordered ocelli and mostly reddish tail. Belly pale yellowish or green. Ten longitudinal rows of ventral scales.

Distribution and habitat Iberia, southern France and north-west Italy. Especially in warm, low and middle altitudes up to 1,000 m, in southern Spain (Sierra Nevada) up to 2,500 m. Inhabits

148

rocky, dry habitats with lots of sun, e.g. verges, forest edges and bushy areas, cultivated areas with sparse vegetation, olive groves and vineyards.

Notes Previously classified as *Lacerta lepida*. Timid, ground-dwelling lizard that climbs well and enjoys sunbathing. When in danger runs noisily into dense vegetation or climbs onto nearby trees. Numerous predators include raptors and the Montpellier Snake. Diet: large insects and other arthropods (e.g. poisonous scorpions), rarely small vertebrates (up to the size of young rabbits). Mating takes place in April and May following a long period of hibernation. Clutches of 5–22 eggs may be laid more than once a year and they hatch after about three months.

Subspecies The nominate subspecies *T. l. lepidus* occupies most of the range, the less contrasting, brownish or greenish grey *T. l. nevadensis* is found in south-east Spain (Sierra

Ocellated Lizard in threat posture

Nevada). The status of two additional subspecies is still debated: *T. l. ibericus* from the north-west of Spain and Portugal and *T. l. oteroi* from the island of Sálvora near La Coruña, Spain.

Similar species Other European lizards are conspicuously smaller. Western Green Lizard has six longitudinal rows of scales on belly. Schreiber's Green Lizard lacks blue ocelli and has only eight rows of belly scales.

Young Ocellated Lizards display a striking pattern of black-bordered white ocelli on the back

Greek Rock Lizard
Hellenolacerta graeca
Lacertid lizards LACERTIDAE

> 23–26 cm
> Greek rock lizard
> Body flattened; back brownish, unstreaked; head long and pointed; cheeks lack large scales

Description Relatively large, elegant, flattened lizard with strikingly long, pointed head and 6–8 cm long body (tail about 2–2.5 times body length). Back glossy grey to bronzy brown, irregularly spotted black (without streaking). Flanks a bit darker but with pale patches. Belly yellow to orange and spotted. Males with blue lower lateral scales. Scaling on head typically has two small scales behind each nostril and no large scales on the cheeks.

Distribution and habitat Southern Greece, only in the mountains of the Peloponnese at altitudes above 400 m. Especially on sunny, not too dry, richly vegetated hills or in open forests. Often found near water.

Notes Climbs well on rocks, walls and trees, and is not particularly shy. Diet: small insects and other invertebrates. Biology hardly known.

Similar species Other lizards have less pointed heads and not so flattened bodies.

Bedriaga's Rock Lizard
Archaeolacerta bedriagae
Lacertid lizards LACERTIDAE

> 20–28 cm
> Typical rock lizards
> Body flattened; back mostly green with reticulated pattern; head long and pointed

Description Large, strong rock lizard with long head that bulges at the cheeks and becomes gradually pointed towards nose. Body 6–8 cm long, strongly flattened. Tail 1.5–2 times as long as body. Back greenish, olive or brownish with striking dark reticulated pattern. Rarely back is plain green or dark brown. Flanks often copper-brown. Belly whitish to reddish, partly spotted. Juveniles have a blue-green tail. Dorsal scales smooth and granular.

Distribution and habitat Only on Corsica, Sardinia and adjacent small islands. Especially in mountains from 500 m–2,700 m; bare boulders; rarely in humid, well-vegetated forests; close to coast in north east of Sardinia.
Notes Agile. Diet: small insects and spiders. Lays 3–6 eggs after 5–6 months hibernation.
Subspecies Nominate subspecies in Corsica, *A. b. paessleri* (pictured) in north Sardinia, *A. b. ferrerae* in north-east Sardinia, *A. b. sardoa* in south Sardinia.

Mosor Rock Lizard
Archaeolacerta mosorensis
Lacertid lizards LACERTIDAE

> 15–22 cm
> Typical rock lizards
> Body dorsally flattened; back grey-brown with dark patches; head long and flat; belly yellowish

Description Mid-sized (body 6–7 cm), very slender, flattened rock lizard with long tail and long pointed head. Upperside variable, glossy grey-brown, olive or brown, mostly with dark patches or barring. Back usually paler than flanks, which can be ventrally bordered by blue patches. Belly plain yellowish to orange. Six pairs of chin scales are typical, with two scales behind each nostril.

Distribution and habitat Very localized in some mountain ranges along coast of south Croatia, Bosnia-Herzegovina and Montenegro. Favours moist, not too sunny hills at altitudes between 600 m and 1,500 m, for example in well-lit deciduous forests, thickets of juniper, close to wells or near crevices on bare limestone karst areas.

Notes One of the rarest European lizards. Very timid, climbs well. Diet: various invertebrates. Lays clutch of 4–6 eggs at the end of July. Juveniles hatch in September.

Sharp-snouted Rock Lizard
Archaeolacerta oxycephala
Lacertid lizards LACERTIDAE

> 18–20 cm
> Typical rock lizards
> Body flattened, grey with pale spots or all black;
> pointed head; blue belly; barred tail

Description Quite small (body length 5.5–6.5 cm), delicate, obviously flattened lizard with long, pointed head. Back reticulated black-brown or brown-grey with pale spots. Can be all dark above in mountains and on some islands. Underside blue. Conspicuous black and blue tail rings. Central pairs of scales on tail underside broader than outer ones. Five pairs of chin shields.
Distribution and habitat Along the Adriatic coast from southern Croatia and Bosnia-Herzegovina to Montenegro, in lowland regions (including many islands off the coast) and coastal mountains up to 1,500 m. Especially on sparsely vegetated, stony habitats like rocky scree, boulders and stone walls.
Notes Timid, agile, good climber. Diet: insects, spiders, other arthropods. Lays 2–4 eggs in June. Hatch after six weeks.
Similar species The very rare Mosor Rock Lizard has six pairs of chin scales and lacks rings on the tail.

Horvath's Rock Lizard
Iberolacerta horvathi
Lacertid lizards LACERTIDAE

> 16–18.5 cm
> Iberian rock lizards
> Body dorsally flattened, grey brown; flanks with well-defined, dark brown band; belly unmarked

Description Small (body size 5.5–6.5 cm), slender, flattened rock lizard with short, blunt head and long tail (up to two times body length). Back pale brown to brown-grey with fine, dark streaks, which often form an indistinct central dorsal line. Flanks with dark brown, well defined band with a wavy fringe. Underside plain yellowish to greenish, with no dark patches. Has five pairs of chin scales.

Distribution and habitat Only in isolated mountain ranges in the north-western Balkans (in Slovenia and Croatia), north-east Italy and southern Austria. Occurs at altitudes of 500–2,000 m, especially in rocky, moist habitats which receive plenty of sun (Plitvice Lakes, Croatia, is a well-known site) or in open deciduous forests.

Notes Diurnal, fast lizard. Diet: spiders and small insects. Between 3–5 eggs are laid in July.

Similar species Common Wall Lizard has a spotted throat.

Aurelios Rock Lizard
Iberolacerta aurelioi
Lacertid lizards LACERTIDAE

> 16–18 cm
> Iberian rock lizards
> Back brown; belly yellowish or greenish, unspotted; found only at high altitudes in the central Pyrenees

Description Small, gracile lizard with a rather flattened body. Body length 5–6 cm, tail about twice body length. Back grey-brown, often with dark patches and pale dorsolateral stripes. Flanks dark brown. Belly plain yellowish to greenish.
Distribution and habitat Only in a small area of the eastern Pyrenees (around Montroig, Pica d'Estats and Coma Pedrosa) at altitudes of 2,100–2,950 m.
Notes Was previously treated as a subspecies of *I. monticola*

(p. 156). Biology hardly known.
Additional species Two similar species, both also recently split, are best identified by range:
■ **Arán Rock Lizard** *Iberolacerta aranica* (distribution in blue) occurs around Valle de Arán at altitudes of 1,900–2,700 m.
■ **Pyrenean Rock Lizard** *Iberolacerta bonnali* (distribution in green) occurs in the central part of Spanish and French Pyrenees (El Portalé, Bonaigua and Lac Bleu de Bigorre) at altitudes of 1,700–3,100 m.

Iberian Rock Lizard
Iberolacerta monticola
Lacertid lizards LACERTIDAE

> 18–23 cm
> Iberian rock lizards
> Green or brown above; belly greenish and spotted;
Portugal only (lower block of red shading on map)

Description Mid-sized (body length 5–8 cm), quite robust, flattened rock lizard with broad flat head, long tail and variable pattern on back. Males green, olive or brownish, with black patches that may form two lateral longitudinal stripes. Flanks generally darker, often with a dark reticulated pattern, some bright spots and blue shoulder patches. Females duller, mostly brown with dark central dorsal line. Belly whitish to yellowish, but during the breeding season (especially in males) strikingly green or yellow-green with fine spots. Juveniles have blue tails.

Distribution and habitat Found only in central Portugal in mountains between 1,100 and 2,000 m. Especially in rocky, sparsely vegetated, quite humid habitats close to timber line or at higher altitudes, for example on boulders or dry stream beds.

Notes Quite resistant to low temperatures. Population density can be quite high. Diet: insects, spiders and other arthropods. Breeding in July, females lay 3–10 eggs under stones, young hatch in 6–8 weeks. The taxonomic status of the genus *Iberolacerta* is still debated. Many herpetologists now regard the geographically separated populations as individual species (eg. *I. galani*, *I. bonnali*, p. 155; *I. cyreni*, p. 158; *I. martinezricai*, p. 159), whereas until recently they were treated as subspecies of *I. monticola*.

Similar species Other lizards within the range lack the green belly. Common Wall Lizard has a spotted throat. Iberian Wall Lizard and Bocage's Wall Lizard are smaller and more delicate with dark lateral streaks stronger than central dorsal streak (if present).

Additional species
■ **León Rock Lizard** *Iberolacerta galani* (range marked as the upper block of red shading on the map) is from north-west Spain.

Iberian Rock Lizard from the Sierra de Estrela, Portugal

(e.g. Covadonga Lakes in the Picos de Europa national park), mainly in mountains but with some populations down to sea level. It was split from Iberian Rock Lizard in 2007 (it was previously treated as *I. m. cantabrica*). It is larger than *I. monticola* and breeding males have a high number of blue ocelli on the shoulders and a plain greenish belly.

Male Iberian Rock Lizard showing typical dark spotted green belly

Spanish Rock Lizard
Iberolacerta cyreni
Lacertid lizards LACERTIDAE

> 18–23 cm
> Iberian rock lizards
> Back mostly green, spotted black; belly plain greenish; only in the mountains of the central Iberian Peninsula

Description Flattened, mid-sized rock lizard; similar to Iberian Rock Lizard (see p. 156), of which it was previously regarded a subspecies. Relatively stout with broad and flat head, long tail and variable patterning on back. Males mostly green or green-brown, rarely brownish, with a pattern of black, partly connected patches. Darker on flanks and often with blue patches on shoulder. Females indistinct, brown or brownish, often with dark central streak on back. Underside plain whitish to greenish or with an indistinct row formed by fine blackish spots on the outer scales of the belly; males often bright green or yellowish green underneath during the breeding season. Juveniles have a blue tail.

Distribution and habitat Only in the Sierra de Guadarrama and Sierra de Gredos in central Spain, at altitudes between 1,300 m and 2,500 m (on Pico Almanzor). Found on rocky overgrown hills; also in humid, sparsely vegetated valleys.

Notes The taxonomic classification of *I. cyreni* is still debated (see p. 157). This species occurs in high densities and is quite resistant to long winters and cool and humid summers. Diet: spiders, small insects such as beetles, flies and ants, as well as other arthropods. Males are highly territorial and each defends a small territory that is occupied by several females. After mating, the females deposit a clutch of 3–8 eggs under stones or in the ground in July or August. The young hatch after about two months.

Subspecies The nominate subspecies is found in the Sierra de Guadarrama, while *I. c. castiliana* inhabits the Sierra de Gredos.

Peña de Francia Rock Lizard

Additional species
■ **Peña de Francia Rock Lizard**
Iberolacerta martinezricai (range in blue) occurs high in the small Sierra de la Peña de Francia. It was previously treated as a subspecies of *I. cyreni*, but is now recognized as a separate species. It is perhaps more closely related to *I. monticola*.

Male of the subspecies castiliana

Common Wall Lizard
Podarcis muralis
Lacertid lizards LACERTIDAE

> 20–25 cm
> Wall lizards
> Back mostly brown; dark central streak on back stronger than lateral stripes; throat clearly spotted

Faetures Mid-sized, elegant, wall lizard with a rather flat body, flat head and long tail. Body size 5–7.5 cm, tail 1.7–2.3 times body length. Collar usually smooth-edged, scales on back slightly keeled. Coloration of back very variable and differs between individuals of a population, and also between different populations. Coloration on the upperside mostly pale or dark brown to grey, sometimes even greenish, with irregular black spots which may merge to form a reticulated pattern; sometimes has no patches at all. Back very often with dark central streak and two parallel longitudinal stripes on the flanks; if additional dark lateral stripes are present, these are usually less developed than the strong dark central streak on the back. On the flanks there is often a dark band of patches surrounded by paler stripes. Especially in Italy green individuals may be found with an intense dark, reticulated pattern across the back. Females are often less contrastingly patterned and the longitudinal streaking is more conspicuous. Sides of the tail have obvious dark and white barring. Underside whitish, yellowish, brownish or (especially among males) reddish to orange (but never greenish or bluish); usually with strong reddish brown or black spots, especially on throat. **Distribution and habitat** Found across Europe, from northern Spain across France and Italy to the southern Balkans. Also on some Mediterranean and Atlantic islands; rare along the Adriatic coast. Does not occur on the Aegean Islands, except for Samothraki. In Britain introduced and established at sites in several counties in southern England. Found in lowlands as well as in hilly regions, in the south up to 2,300 m. Especially favours dry and

Male Common Wall Lizard from the Italian Abruzzo region

sunny habitats like vineyards, quarries, dry meadows, embankments, dry-stone walls, ruins or forest edges. In the southern part of the range it is found mostly at higher altitudes in quite humid and shady places, for example in open deciduous forests or on rocky hills with plenty of crevices. **Notes** Warmth-loving, diurnal, agile species that often climbs

Common Wall Lizard from the Spanish Pyrenees

Common Wall Lizard of the sub-species albanica *from Albania*

on stone walls. Diet: insects, spiders and other arthropods, also slugs, caterpillars and worms. The tail can be shed, like in almost all species of lizards (autotomy). Mating takes place in March–April, depending on habitat and location. The males occupy small territories that they defend against intruders; as the opponents fight and bite each

Male Common Wall Lizard of the nominate subspecies

other they often lose all their typical wariness. In May–July the females lay clutches of 2–12 soft-shelled eggs in the ground or under stones. If climatic conditions are good, 2–3 clutches are laid per year, in cooler regions just one. After 2–3 months, 6-cm long young hatch.

Subspecies About 30 subspecies have been described (with 18 in Italy alone), which are slightly different in patterning and scaling, but cannot always unanimously be assigned to one subspecies and are thus still debated. At the moment nine subspecies are recognized: the nominate subspecies *P. m. muralis* from Austria, eastern central Europe and neighbouring parts of Italy (also introduced into southern England and Ohio, USA); *P. m. nigriventris* from western Italy from Liguria and Tuscany along coast to Naples (also introduced in

Passau and Dresden, Germany); *P. m. albanica* from the southern Balkans (Albania, Greece, Bulgaria) and adjacent Turkey; *P. m. merremia* from central and eastern Spain, southern France and western Liguria (Italy), western Switzerland and the Rhine valley (as far as Bonn); *P. m. brogniardi* from north-west Spain, high altitudes of the Pyrenees, western and central France, Belgium, the Netherlands and north-west Germany; *P. m. colosii* from Elba (Italy) and the adjacent mainland; *P. m. maculiventris* from northern Italy, Tessin (southern Switzerland), Slovenia and north-west Croatia (also introduced to British Columbia, Canada); *P. m. breviceps* from the Calabrian mountains (south Italy); *P. m. brueggemanni* from north-west Italy.
Similar species Common Lizard has a serrated collar. Iberian

Mating pair of the Common Wall Lizards of the subspecies albanica

Wall Lizard has a more weakly spotted throat and no dark dorsal stripe (if it is present, then it is always weaker than the lateral stripes); if both species occur together, *P. hispanica* is on sun-exposed, rocky ground and *P. muralis* on slightly overgrown ground. Italian Wall Lizard has a plain underside. Aegean Wall Lizard is less flattened and only a weak (or no) dark dorsal stripe.

Green-backed male Common Wall Lizard of the subspecies nigriventris *from Italy*

Iberian Wall Lizard
Podarcis hispanica
Lacertid lizards LACERTIDAE

> 16–20 cm
> Wall lizards
> Back brown; dark lateral stripes stronger than the often lacking central dorsal stripe; throat spotted

Description Mid-sized (body 5–7 cm), gracile and flattened with a pointed snout. Back brown, grey or greenish, flanks dark with variable pale and dark patches. Females often have four pale, dark-edged lateral streaks. Dark dorsal streak (if present) weaker than lateral stripes. Belly white to reddish, throat spotted black. **Distribution and habitat** Iberian Peninsula, south-west France and north-west Africa, from sea level to 3,480 m (Sierra Nevada). Often in dry, steep, rocky areas.

Notes Climbs very well. Clutch consists of 2–5 eggs. **Subspecies** Nominate subspecies widely distributed, *P. h. vaucherii* (sometimes treated as a separate species) in south-west Spain and northern Africa, and *P. h. cebennensis* in the Pyrenees and southern France. **Additional species** ■ **Columbretes Wall Lizard** *Podarcis atrata* (range in blue) is similar. Only on the Columbretes Islands. Previously treated as a subspecies of *P. hispanica*.

Bocage's Wall Lizard
Podarcis bocagei
Lacertid lizards LACERTIDAE

> 16–20 cm
> Wall lizards
> Back green; flanks orange-brown, spotted black;
  Belly dark, patchy; north-west Iberian Peninsula only

Description Similar to Iberian Wall Lizard (of which it was regarded a subspecies until recently) but less flattened, more robust and the females are less longitudinally striped. Body length 5–7 cm, tail about twice that length. Back of males mostly green. Flanks orange-brown with a dark reticulated pattern, often bordered by bright longitudinal stripes; sometimes with blue patch on shoulder. Belly whitish to orange, intensely patterned with dark spots.

Distribution and habitat North-west Iberian Peninsula in lowland and hilly countryside, sea level up to 1,900 m. Dry, rocky, partly vegetated slopes.
Notes Less adept at climbing than *P. hispanica*.
Additional species
■ **Carbonell's Wall Lizard**
Podarcis carbonelli (range in blue) is gracile and the male has green flanks. Nominate subspecies on mainland and *P. c. berlengensis* on the Berlengas Islands. Only recently split from *P. bocagei*.

165

Ibiza Wall Lizard
Podarcis pityusensis
Lacertid lizards LACERTIDAE

> 20–25 cm
> Wall lizards
> Back often green with rows of blackish spots; scales slightly keeled; throat spotted; Balearic Islands

Description Mid-sized (body length 6–9 cm), robust species with short head and long tail. Body not flattened. Coloration of back very variable, mostly pale green, blue-green, brownish or yellowish but also grey to almost black; often with small black spots which can merge into three longitudinal rows bordered by bright lines, especially on the central row. Flanks have black patches, often in reticulated patterns, and blue patches on lower lateral scales.

Underside variable with whitish, pale grey, yellow, bluish, greenish, orange or reddish markings, especially laterally. Throat spotted dark. On some islands individuals with a very dark, almost plain black upperside and blue belly occur. Back scales longish and hexagonal, and slightly keeled. **Distribution and habitat** Found only on Ibiza, Formentera and their numerous adjacent islets. Was introduced to Mallorca and there was a small population in the centre of Barcelona (Plaza de las Glorias) around 1990 . On Ibiza common in all habitats from sea level to the highest point of the island (475 m). Often found around ruins, bushes and walls or on stony, fallow land. On the small islets it inhabits bare rocks.

Notes Diurnal lizard that occurs in high densities, not very timid. Diet: insects and other arthropods as well as ripe fruits.

Subspecies Due to the scattered distribution on numerous small isolated islets, many very different-looking variants evolved (even dwarf and giant forms), so that more than 40 subspecies have been described, 22 of which are currently recognized. Among them are *P. p. pityusensis* from Ibiza and *P. p. formenterae* from the neighbouring Formentera.

Additional species

■ **Lilford's Wall Lizard** *Podarcis lilfordi* (range in blue) is closely related and is split into about 25 subspecies. It inhabits the neighbouring islets of the Balearic Islands (the Cabrera Islands as well as many islets

Blue-tinged male Ibiza Wall Lizard (P. p. vedrae) from Illa de Vedrà

around Mallorca and Menorca, but not both main islands except for a few introduced populations). Distinguished by fine, smooth scaling of the back (rougher and weakly keeled in Ibiza Wall Lizard).

Two additional species of wall lizards (that do not belong to the genus *Podarcis* according to the latest research) have been introduced into Europe:

■ **Moroccan Rock Lizard** *Teira perspicillata* is a small species from north-west Africa, which prefers a rocky habitat and is introduced to Menorca.

■ **Madeira Lizard** *Teira dugesii* from Madeira is introduced around Lisbon harbour (Portugal).

Lilford's Wall Lizard from the islet of Colon near Menorca

Tyrrhenian Wall Lizard
Podarcis tiliguerta
Lacertid lizards LACERTIDAE

> 20–25 cm
> Wall Lizards
> Back mostly greenish brown with longitudinal band of patches; throat spotted; only Corsica and Sardinia

Description Mid-sized wall lizard with a strong, stout head. Unflattened body up to 6.5 cm long, tail twice that length. Upperside variable, brown to grey-brown, males greenish with bright longitudinal stripes on flanks and black patches that often form longitudinal stripes and a reticulated pattern. Flanks darker, spotted black or sometimes blue. Belly whitish, yellowish, orange or red, spotted black on throat. Supratemporals curved downwards at the sides.

Distribution and habitat Only on Corsica and Sardinia (including adjacent islands), in lowlands and mountains up to 1,800 m. Especially favours dry, stony, sometimes overgrown mountain slopes or embankments.
Notes Diurnal, mainly ground-dwelling species. Mating in April after five-month hibernation period. Clutches of 6–12 eggs.
Subspecies Apart from nominate subspecies, ten subspecies of unclear status have been described from surrounding islets.

168

Sicilian Wall Lizard
Podarcis wagleriana
Lacertid lizards LACERTIDAE

> 22–25 cm
> Wall lizards
> Back mostly green with bright lateral stripes and rows of black patches; throat spotted; Sicily only

Description Mid-sized, slender species with relatively pointed and flat head. Slightly flattened body up to 7.5 cm, tail twice that length. Back green (females may be olive-brown), mostly with pale dorsal stripes and black patches that form 1–3 longitudinal lines. Flanks with black and some blue patches, but some may lack patterning. Belly whitish, orange-red in males; throat spotted black.

Distribution and habitat Sicily (except for the north-east) and the Egedi Islands, especially in lowland and hilly country up to 1,600 m. Dry, sunny places with grass and bushes, even towns.

Notes Diurnal, ground-dwelling, rarely climbs. Lays 4–6 eggs.

Subspecies *P. w. marettimensis* from the Egadi Islands and the nominate subspecies elsewhere.

Additional species

■ **Maltese Wall Lizard** *Podarcis filfolensis* (distribution in blue) is closely related and split into five subspecies living on Malta, Gozo and neighbouring islands.

Italian Wall Lizard
Podarcis sicula
Lacertid lizards LACERTIDAE

> 20–25 cm
> Wall lizards
> Back often green, spotted black with brown dorsal stripe; throat and belly plain; keeled dorsal scales

Description Relatively large and strong wall lizard with flattened, longish head and long tail. Body 6–9 cm (individuals in the north of the range generally a little smaller than those in the south), tail over twice that length. Coloration of back very variable, bright green, olive, yellowish, grey or brownish, usually with black patches that form reticulated or streaked patterns. Often individuals from the north (rarely from the south) have pale dorsolateral stripes, and a brown central dorsal stripe interspersed with black patches. Pale olive-green or very dark plain individuals sometimes occur. Males especially may show a blue patch above front legs and black and blue spotted lower lateral scales. Females brownish, usually with more obvious longitudinal lines. Belly white-, yellow-, red- or greenish, sometimes ultramarine blue, never spotted, even on throat. Dorsal scales keeled.
Distribution and habitat Italy and northern Adriatic coast

with adjacent islands (south to Dubrovnik), Corsica, Sardinia, Sicily and neighbouring small islets of the Tyrrhenian Sea. Also introduced to Spain (Almeria, Santander, Menorca), northern Africa, Portugal (Lisbon),Turkey (Istanbul) and the USA (Philadelphia, New York). Found in lowlands and mountainous regions up to 1,800 m (near Mt Etna, Sicily). Common on dry boulders surrounded by vegetation, by gravel pits and on farmland, pastures, embankments, dry stone walls and verges.

Notes Diurnal, ground-dwelling and very adaptive. One of the most common lizards of southern Europe. May hybridize with other wall lizards, e.g. *P. melisellensis*, *P. wagleriana*, *P. tiliguerta* and *P. raffonei*. Mating in April and May, clutches contain 6–12 oval eggs.

Subspecies To date 91 subspecies (especially island forms) have been described, about 45 of which are considered valid today. Many others are still debated. The most widespread are the nominate *P. s. sicula* from southern Italy and *P. s. campestris* from central and northern of Italy, Corsica and the Dalmatian coast.

Similar species Bedriaga's Rock Lizard has smooth, slightly granular dorsal scales. Tyrrhenian Wall Lizard has a spotted throat. Sicilian Wall Lizard has a bold bright stripe on the back. Dalmatian Wall Lizard is small

Male P. s. campestris *from Lago Trasimeno, Italy*

and gracile with a short head.

Additional species

■ **Aeolian Wall Lizard** *Podarcis raffonei* (range in blue) lives only in the Aeolian Islands; the nominate subspecies (originally described as a subspecies of *P. sicula*) on Strombolicchio and *P. r. antoninoi* (originally described as a subspecies of *P. wagleriana*) on Vulcano.

Italian Wall Lizard of the nominate subspecies from Sicily

Dalmatian Wall Lizard
Podarcis melisellensis
Lacertid lizards LACERTIDAE

> 15–19 cm
> Wall lizards
> Back green or brown, often with bright lateral
  stripes; throat unspotted; northern Adriatic coast

Description A small, delicate species (body size 5–6.5 cm, tail twice that length) with a short head. Back green, olive or brown, with two bright dorsolateral stripes and a dark central band or no pattern at all. Flanks have dark and pale rows of patches and blue lower lateral scales. Males have a blue shoulder patch. Belly plain orange-red in breeding males, otherwise whitish.

Distribution and habitat Adriatic coast from north-east Italy to northern Albania and adjacent islands. Found in dry lowland and hilly country up to 1,300 m, especially on bare ground, embankments and stone walls.

Notes Common, diurnal and ground-dwelling. Males form territories during breeding.

Subspecies The nominate *P. m. melisellensis* inhabits the island of Melisello (Brusnik), *P. m. fiumana* is found on the Dalmatian mainland. The status of 18 additional subspecies from the adjacent islands is still under discussion.

Balkan Wall Lizard
Podarcis taurica
Lacertid lizards LACERTIDAE

> 20–24 cm
> Wall lizards
> Centre of back green; pale lateral stripes; flanks brown, spotted; serrated collar; throat plain

Description Mid-sized, robust and slightly flattened with a short head. Body 6–8 cm, tail twice that length. Centre back green, unspotted, mostly with two bright dorsolateral stripes, brown flanks spotted black; rarely with no pattern. Often has blue spots on lower lateral scales and blue patches on shoulder. Belly and throat plain whitish to orange-red. Collar serrated.
Distribution and habitat Southern Balkans north to Hungary. Lowlands and hilly country up to 1,200 m, not too dry cultivated land, grassland, embankments, walls, vegetated dunes.
Notes Ground-dwelling species that rarely climbs.
Subspecies *P. t. ionica* in Albania and west Greece, *P. t. thasopulae* in Thasopula, nominate subspecies in rest of range.
Additional species
■ **Skyros Wall Lizard** *Podarcis gaigeae* (range in blue) is very similar. Only on Skyros islands. Previously regarded a subspecies of Erhard's Wall Lizard, p. 174.

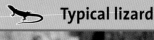

Erhard's Wall Lizard
Podarcis erhardii
Lacertid lizards LACERTIDAE

> 18–22 cm
> Wall lizards
> Back mostly brown with light and dark dorsolateral stripes; southern Balkans and many Aegean islands

Description Mid-sized, relatively slender species with a slightly flattened body and broad head. Body up to 6–7 cm, tail about twice that length. Collar not usually serrated, scales unkeeled. Back brownish, copper or sandy coloured, mostly with pale and dark dorsolateral stripes, wider and stronger than the central back stripe which is often lacking on individuals from the mainland. Males often show a reticulated pattern on the flanks and blue lower lateral scales.

Island individuals often larger (body up to 8 cm), with weakly keeled scales, slightly serrated collar and a more variable coloration. Back can be brown, grey or green, often with dark reticulated pattern but can be plain. Belly whitish or yellowish, orange to red in males, throat dark, can be spotted.

Distribution and habitat Southern Balkan peninsula, from Albania to southern Greece (except for wide parts of the Peloponnese) including many Aegean islands, where this species is often the only lizard. Lowland and hilly countryside up to 1,000 m, especially in rocky, dry places with low and bushy vegetation or vegetated sand dunes, rarely on embankments and walls.

Notes Ground dwelling, diurnal, rarely climbs. Diet: small insects and arthropods. Mating mostly March and April. Small clutch of 2–4 eggs laid in summer.

Subspecies Nominate subspecies *P. e. erhardii* inhabits the Aegean islands of Serifos and Sifnos. Three additional subspecies are described from the mainland: *P. e. livadiaca* in central Greece and the Peloponnese, *P. t. thessalica* in Thessalia and *P. e. riveti* from the rest of the range. An additional 23 subspecies live on Crete and various Aegean islands, e.g. *P. e. naxensis* on the Cyclades islands of Naxos, Ios and Thira.

Similar species Balkan Wall Lizard and Skyros Wall Lizard have a strongly serrated collar. Common Wall Lizard has a strongly spotted throat and a dark central streak on back which is always more prominent than the lateral stripes.

Additional species

■ **Milos Wall Lizard** *Podarcis milensis* (range in blue). Males have a black throat. Pale spots on front of body. The three subspecies inhabit the Cyclades islands of Milos, Antimilos and Falkonera. Like the Skyros Wall

Male of the subspecies riveti

Lizard, they were previously treated subspecies of Erhard's Lizard or Balkan Wall Lizard (see p. 173).

■ **Anatolian Wall Lizard** *Podarcis anatolica* (range in green). Wall lizard populations on some islands off the Turkish coast may be this 'species', but the systematic classification is still unclear. They occur on Samos, Rhodes and adjacent islands, Ikaria and Cyprus, but further research may result in splits and reclassifications.

P. e. riveti *from Bulgaria*

Peloponnese Wall Lizard
Podarcis peloponnesiaca
Lacertid lizards LACERTIDAE

> 22–27 cm
> Wall lizards
> Back brownish, often with bright longitudinal stripes; flanks often blue; throat plain; only in Peloponnese

Description Largest wall lizard, robust, strong head, body 6–8.5 cm (tail about twice that length). Upperside brownish, greenish in males, often with distinct longitudinal stripes. Females have contrasting pale and dark brown pattern, usually with two to four bright dorsolateral lines and dark longitudinal stripes formed by dark spots, always better developed than central stripe (but some have no markings). Often has blue lower lateral scales and prominent blue patches on flanks. Underside plain whitish, orange or red. Collar smooth.

Distribution and habitat Greek Peloponnese, cultivated areas in lowlands and hilly country up to 1,600 m, especially dry, stony, barren land, olive groves, ruins and embankments.

Notes Diurnal, ground-dwelling.

Subspecies *P. p. peloponnesiaca* from the southern Peloponnese. Two controversial subspecies from north (*P. p. lais*) and north-east (*P. p. thais*) are described.

Meadow Lizard
Darevskia praticola
Lacertid lizards LACERTIDAE

> 14–16 cm
> Meadow lizard
> Scales on back larger in centre than on the sides; back pale brown, flanks dark brown; throat plain

Description Small, delicate, body 5–6 cm, tail 1.5–2 times that length. Short legs and rough, keeled scaling. Scales smaller and narrower on the sides than in the centre. Quite uniformly coloured above with wide, pale to red-brown dorsal stripe bordered by blackish spots. Flanks dark brown bordered by band of bright patches on lower side. Belly plain whitish to greenish-yellow. Collar serrated. **Distribution and habitat** Patchily distributed in eastern Balkans (parts of Serbia, Romania, Bulgaria and Greece). Lowlands up to 800 m, in moist, shady deciduous forests or on verges and well-vegetated pastures. **Notes** Diurnal, quiet, mainly ground-dwelling, accomplished climber. Clutch size 3–6 eggs. **Subspecies** The nominate subspecies *D. p. praticola* in central Caucasus, *D. p. pontica* in Europe and western Caucasus. **Similar species** The dorsal and lateral scales of the Common Lizard are of the same size.

177

Common Lizard
Zootoca vivipara
Lacertid Lizards LACERTIDAE

> 12–18 cm
> Common lizard
> Back brown; dark central dorsal line and flanks; small, keeled scales on back; serrated collar

Description Slender, small to mid-sized lizard appears quite robust due to short legs and small, slightly flattened head. Tail less than twice as long as the 6–7 cm long body. Keeled scales on back. Upperside brown, grey or bronze, sometimes completely black. Dark central dorsal line can be spotted, especially among males. Also has rows of small spots. Flanks, especially in females, have a wide, dark brown longitudinal stripe with bright borders above and below. Underside whitish to cream. On females plain or with a few black patches, on male yellowish to orange and more prominently spotted. Collar serrated.

Distribution and habitat Inhabits wider range and more northerly areas than any other reptile species (beyond 70° north, distance from west to east edges

of its distribution more than 11,000 km). Found from Ireland and north-west Spain across Europe and Asia east to Siberia, Sakhalin and Hokkaido. Favours moist, cool lowland areas, also in middle and high altitudes, in the Pyrenees up to 2,400 m, in the Alps up to 3,000 m. In some habitats can be very common, e.g. gravel pits, pastures, mountain meadows, heaths, forest edges, embankments and clearings.
Notes Diurnal, quite sedentary with slow, deliberate movements. May shed tail if in danger (autotomy) and will even escape into water. Diet: small insects, spiders, centipedes, isopods. In Britain mating takes place from April to June. Viviparous. Young hatch three months later from 3–12 transparent eggs within female's body. In south-west and southern Europe, egg laying populations are also known which deposit two clutches between June and August. These are regarded as a separate subspecies (see below). Hibernation generally between October and February.
Subspecies The classification of the different subspecies is still a subject of controversy. Apart from *Z. v. sachalinensis* from eastern Siberia and the nominate subspecies *Z. v. vivipara*, which inhabits a wide part of the range, only two additional subspecies are recognized: *Z. v. pannonica* from the lowlands of eastern

Common Lizard from the Cantabrian Mountains, Spain

Slovakia and the egg-laying *Z. v. carniolica* from Slovenia and adjacent areas. Egg-laying Common Lizards in Spain may also form a separate subspecies.
Similar species Other lizards with serrated collars. In Britain the Sand Lizard and the introduced Common Wall Lizard both have a larger head and longer legs. Sand Lizards have a characteristic row of small scales in the centre of the back. In south-east Europe the scales of the Meadow Lizard are much larger on centre back than on the sides.

Pregnant female Common Lizard

Snake-eyed Lizard
Ophisops elegans
Lacertid lizards LACERTIDAE

> 15–19 cm
> Snake-eyed lizard
> Back brownish with four bright lateral stripes;
  eyelids cannot move and are transparent; lacks collar

Description Small slender lizard. Body 5–6 cm, tail 1.5–2 times that length. Transparent eyelids grown together to form 'glasses' (eye does not close when gently touched). Back scales large and keeled. Upperside brownish with four narrow, bright, dorso-lateral stripes bordered by lines of black patches. Flanks often greenish or yellowish with a reticulated pattern. Belly plain whitish; male's throat greenish to bluish. No collar.

Distribution and habitat The Middle East, in Europe only in the very south-east and on some Aegean Islands (e.g. Rhodes, Thassos). Found from sea level up to 800 m in open, dry, very sunny areas; stony hills, meadows or olive groves.

Notes Agile, diurnal, ground-dwelling. Clutch size 2–6 eggs.

Subspecies Europe is inhabited by *O. e. macrodactylus* (originally *O. e. ehrenbergii*), with *O. e. schlueteri* on Cyprus. The nominate subspecies and other subspecies live outside Europe.

Large Psammodromus
Psammodromus algirus
Lacertid lizards LACERTIDAE

> 25–30 cm
> Psammodromus lizards
> Large scales which overlap, keeled; back brown with four bright longitudinal stripes; no collar

Description Large, robust lizard, appearing quite slender due to the long tail which is 2–3 times the body length (7–9 cm). Large dorsal scales, pointed and overlapping, strongly keeled (even behind the ears). Upperside glossy pale to dark brown with two bright lateral longitudinal stripes bordered by dark lines, often with dark central streak. Males with blue shoulder patches and orange throat and cheeks in mating season. Underside whitish to green. No collar.

Distribution and habitat Across most of the Iberian Peninsula and southern France west of the Rhone; also in north-west Africa. Occurs from sea level to 2,600 m (Sierra Nevada, Spain), common in open countryside with plenty of bushes, or in open forests with dense undergrowth, particularly alongside streams on large stones.

Notes See Spanish Psammodromus (p. 182), can also call, climbs better than that species. Clutch size 2–11 eggs.

Spanish Psammodromus
Psammodromus hispanicus
Lacertid lizards LACERTIDAE

> 13–15 cm
> Psammodromus lizards
> Scales keeled and overlapping; back with 4–6 bright lines, barred black; hint of a collar

Description Small, slender lizard with a slightly flattened body and small head; tail not more than twice the length of the small body, which is 4–5 cm long. Back scales strongly keeled and overlapped, but this may be hard to discern in the field due to the small size of the lizard. Upper side with variable tones of brown, grey, olive or ochre, mostly with a characteristic longitudinal pattern of 4–6 whitish or yellowish stripes, which may be broken, as well as interspersed black barring ('segmented' appearance); rarely uniform grey brown. Flanks of males often green during breeding season. Underside whitish, slightly greenish or reddish, plain or only slightly spotted. Only a hint of a collar on the sides, scales right behind the ear granular and unkeeled.
Distribution and habitat Most parts of the Iberian Peninsula, southern France (Mediterranean coast, lower Rhone Valley), more patchily distributed than

the Large Psammodromus, but may be quite common in suitable habitat. Found in lowland areas and hilly countryside, from sea level up to 1,700 m (Sierra de Guadarrama and Sierra Nevada). Dry, semi-open habitat with patches of dense, low bushes, cultivated areas and fallow land or in sandy dunes.

Notes Diurnal, very agile, ground-dwelling. When disturbed runs at high speed across open areas in a very characteristic way to find a hiding place in the sand or grass. May utter squeaking sounds during the breeding season or when captured. Diet: small insects, spiders and other arthropods. Clutches consist of 3–6 eggs, which are deposited in the earth once or twice a year. The juveniles hatch between June and August.

Subspecies The nominate

Individual with characteristic 'segmented' pattern on back

subspecies *P. h. hispanicus* inhabits Portugal and central and southern parts of Spain, while *P. h. edwardsianus* occurs in north-east Spain and south-west France.

Similar species Large Psammodromus is much larger, with a proportionally longer tail and keeled scales behind the ear. Other lizards have a more obvious collar.

Spanish Psammodromus from Alicante, southern Spain

Spiny-footed Lizard
Acanthodactylus erythrurus
Lacertid lizards LACERTIDAE

> 20–23 cm
> Spiny-footed lizards
> Back brown with 8–10 bright lines; no occipital scale; two supraoculars

Description Mid-sized, rather stout lizard with short, upright head and pointed snout. Tail about twice the length of the 6–7.5 cm-long body. Lower back scales keeled and slightly enlarged. Upperside has a variable pattern, mostly brown, grey brown or ochre with 8–10 bright lines (may be indistinct or broken). In between are bright and dark patches, which form a bar-like pattern; rarely (especially in the south) almost uniform plain grey. Legs have bright, round, patches bordered black. Underside plain whitish grey. Juveniles conspicuously striped black and white with reddish hind legs and tail. 'Spiny scales' on the sides of fingers and toes (thus the name Spiny-footed Lizard), which are generally quite weakly developed in this species but which are particularly well developed on the fourth toe. Collar present and has distinctive head-scaling: only two supraoculars above eyes; no occipital scale;

subocular scale does not reach the upper lip or barely does so.

Distribution and habitat Spain and southern and central Portugal, north-west Africa. Occurs in lower and middle altitudes; does not exceed 1,800 m in the Spanish Sierra Nevada. Common in dry, hot places with sandy ground, especially in dunes close to the sea (the spiny feet act like snowshoes to enable faster movement), but also on rocky ground with sparse vegetation further inland.

Notes Diurnal, ground-dwelling, not shy and very agile. Escapes by running in a distinctive way across open ground with tail slightly raised. Only runs short distances. If relaxed, the front part of the body is raised. The species is very vigilant. Juveniles move their tails slowly back and forth when sunbathing. Diet: insects and spiders. During the breeding season the males are very aggressive and occupy small territories, which they strongly defend against intruders. The females deposit a clutch of 4–6 eggs in the ground once or twice a year. Adults will hibernate but juveniles usually do not.

Subspecies The nominate *A. e. erythrurus* lives in Portugal and Spain. Two additional subspecies are found in Africa.

Similar species Other European collared lizards do not have spiny scales on the fingers and toes, most of them have a striking occipital shield and four supraoculars.

Individual with typical coloration

Dalmatian Algyroides
Algyroides nigropunctatus
Lacertid lizards LACERTIDAE

> 18–21 cm
> Algyroides lizards
> Scales keeled, larger in the centre of back than on the flanks; males have blue neck; flanks orange

Description Mid-sized, slender lizard with long and high head. Tail about twice the size of the 6–7 cm-long body. Back scales not pointed, strongly keeled, overlapping and much larger than the keeled lateral scales. Upperside dark, mostly grey or reddish brown with small black spots, but without larger patches. Males typically have deep blue on the neck and often on the sides of the head, as well as a bright orange underside and partly orange flanks. Some lower ventral scales are also mostly blue. Underside of the females is plain yellowish to greenish white. Well-developed collar is strongly serrated.

Distribution and habitat From north-east Italy and Istria along the Adriatic coast (also on many islands) in scattered populations to Albania and north-west Greece (Epirus Mountains, as well as some Ionian islands). Usually in places rich in vegetation and with some shade; for example in olive groves, leaf litter under bushes, along streams, walls and hedges, and also close to human settlements and fields.

Notes Diurnal, ground-dwelling lizard that likes to stay hidden. Diet: insects, spiders and other arthropods, but also small worms and caterpillars. Males defend a territory during the April and May mating season. The females deposit small clutches consisting of 2–3 eggs once or twice a year in the ground.

Subspecies The nominate subspecies *A. n. nigropunctatus* inhabits almost all of the range including the Greek mainland and the adjacent islands of Corfu, Paxos und Lefkas. Another subspecies, *A. n. kephallithacius,* has been described from the Greek

This species often remains under cover in dense vegetation for long periods

islands of Cephalonia and Ithaca.

Similar species Most other species of algyroides have small, but not keeled, or only weakly keeled dorsal scales that lie side by side. Greek Algyroides has pointed dorsal scales that are as large as the lateral scales.

Adult male Dalmatian Algyroides of the nominate race

Greek Algyroides
Algyroides moreoticus
Lacertid lizards LACERTIDAE

> 10–13 cm
> Algyroides lizards
> Keeled scales, pointed, and equally sized on the back and flanks; centre of back uniform brown

Description Small, delicate lizard with a tail about 1.5–2 times as long as the 4–5 cm-long body. Back scales overlap like roof-tiles, keeled, pointed and as large as scales on flanks. Centre of back in males is uniform red-dish to dark brown, mostly with two yellowish dorsolateral stripes and greenish yellow flanks with dark marbling. Females all brown. Underside whitish to greenish yellow, mostly spotted dark. Legs often spotted white. Collar strongly serrated.

Distribution and habitat Greece (Peloponnese) and on the three Ionian islands: Cephalonia, Itha-ca and Zakynthos. Especially in north-facing, shady habitats, for example, on moist boulders rich in vegetation, clearings and close to streams. Also favours bushes and hedges.

Notes Diurnal, ground-dwelling lizard with secretive habits. Diet: mainly small insects and spiders.

Similar species See Dalmatian Algyroides, p. 186.

Pygmy Algyroides
Algyroides fitzingeri
Lacertid lizards LACERTIDAE

> 12–13 cm
> Algyroides lizards
> Keeled and pointed scales; small delicate body; back-brown with small spots

Description Very small and delicate lizard with a small, flattened head and strong tail, about twice as long as the 3.5–4 cm-long body). Back scales quite large, keeled, pointed, and overlap like roof tiles. Back uniform grey to dark brown or dark olive, mostly with dark spots, which may form a fine central line on back. Underside whitish, yellowish or orange, throat spotted. Serrated collar.

Distribution and habitat Only in Corsica and Sardinia, at altitudes up to 1,100 m, especially on moist boulders, slopes with some shade, in undergrowth or on walls and old trees.

Notes Very secretive, diurnal and ground-dwelling lizard. Lays clutches of 2–4 eggs.

Additional species

■ **Spanish Algyroides** *Algyroides marchi* (distribution in blue) only inhabits three small mountain ranges in south-east Spain at altitudes of 700–1,700 m (Sierra de Cazorla, Sierra de Segura and Sierra de Alcaraz).

Bedriaga's Skink
Chalcides bedriagai
Skinks SCINCIDAE

> 13–17 cm
> Skinks
> Small, elongated; limbs short; back has small, bright patches bordered dark brown

Description A small, elongated skink with small head and short five-toed limbs. Tail about as long as the 6–8.5 cm-long body. Back with smooth, glossy brown, grey, olive or yellow-brown scales, with small, scattered white ocelli bordered dark brown. Indistinct stripe along centre of the back and flanks. Underside pale and unmarked.

Distribution and habitat Iberian Peninsula, patchily distributed in lower and middle altitudes (rarely up to 1,800 m). Found in open and dry areas with little vegetation and in well-vegetated, moist habitats.

Notes Diurnal. Leads a secretive life in loose ground or in leaf litter. Viviparous and gives birth to between 1–4 young in a year.

Subspecies The nominate subspecies *C. b. bedriagai* is found in the south of the range and *C. b. cobosi* in the east of the Iberian Peninsula. *C. b. pistaciae* from the west has been split by some authorities as a separate species (Cylindrical Skink).

Ocellated Skink
Chalcides ocellatus
Skinks SCINCIDAE

> 25–30 cm
> Skinks
> Large, elongated; short legs; back brown with white ocelli, which are bordered black

Description Large, elongated skink with small, pointed head and short operating limbs. Tail short with a sharp tip, between a third and a half as long as the body. Back glossy yellow-brown to grey with longish, white ocelli bordered black, which are sometimes ordered crosswise. Underside plain whitish.

Distribution and habitat Peloponnese (Greece) and Mediterranean islands like Sardinia, Sicily and Crete at altitudes up to 1,200 m; also north Africa and south-west Asia. Varied but mostly dry habitats, such as sand dunes, cultivated fields, olive groves and ruins.

Notes Agile lizard that lives in loose sand and under stones. Females give birth to 5–15 young.

Subspecies The nominate subspecies is found in Greece and *C. o. tiligugu* in Italy.

Additional species
■ **Levant Skink** *Mabuya aurata* (distribution in blue) only occurs on a few islands off the Turkish coast (for example, Rhodes).

Italian Three-toed Skink
Chalcides chalcides
Skinks SCINCIDAE

> 35–48 cm
> Skinks
> Body snake-like, glossy with longitudinal streaking; very short legs with three toes

Description Relatively large, elongated, snake-like skink with a small head that merges into the body and with tiny, three-toed limbs which are functional but which can easily be over-looked. Unbroken tail long with a pointed tip; reaches at least half or two thirds of the body length. Upperside glossy bronze or olive- or yellowish-brown. Pattern variable, often has two bright longitudinal stripes that are partially bordered dark and which enclose a slightly darker stripe on the back; often with 4–6 dark longitudinal stripes or no pattern at all. Underside plain whitish and brighter than the upperside.

Distribution and habitat Italy and adjacent Mediterranean islands (e.g. Sardinia, Sicily, Malta, but not Corsica), in lowland and hilly countryside up to 1,600 m (Sicily). Very specialized species which favours habitats with dense and low grassy vegetation, especially on sandy, sunny, but quite moist ground, for example fallow land, grassy edges of trails and dry and moist pastures. Avoids forests and tall and dense vegetation.

Notes Ground-dwelling, very agile lizard that hunts for prey in low vegetation, especially in the mornings. Diet: spiders, insect larvae, small insects and isopods. Prey is pursued quickly by pressing the limbs into the flanks and shooting rapidly through the undergrowth. Mating takes place from mid-March after a hibernation period of about five months. The males have fierce territorial fights in which their tails may become damaged. Females are viviparous and give birth to 4–9 young.

Additional species

■ **Western Three-toed Skink**
Chalcides striatus (distribution in blue) replaces its Italian relative in the Iberian Peninsula,

Italian Three-toed Skink from Sicily

southern France and adjacent north-west Italy. Found up to 1,800 m, usually distinguished by 9–13 dark longitudinal stripes and the second finger about as long as the third (in the Italian Three-toed Skink the second finger is longer than the third). Only recently split from *C. chalcides* as a separate species.

Western Three-toed Skink

Snake-eyed Skink

Ablepharus kitaibelii
Skinks SCINCIDAE

> 10–13.5 cm
> Skinks
> Very small and gracile; legs short; back glossy brown; eyelids rigid, transparent

Description Diminutive, delicate, elongated skink with short but functioning legs and a quite broad tail. Eyelids grown together to form transparent 'glasses' (hence 'snake-eyed'). Back glossy brown to olive, plain or with dark spots in longitudinal lines. Flanks dark brown. Underside plain greyish.

Distribution and habitat Eastern and south-east Europe, where found in sunny, warm places at low altitudes. Dry meadows, fallow land, olive groves, bright deciduous forests.

Notes Diurnal and agile. Lives secretively in leaf litter or under stones. Clutch consisting of 2–4 eggs is deposited in the ground; young hatch after two months.

Subspecies The nominate subspecies from the Peloponnese, Cyclades and southern Sporades, *A. k. fitzingeri* from the Czech Republic and Hungary, *A. k. stepaneki* from the Balkan Peninsula and *A. k. fabichi* from some east Aegean islands (e.g. Kasos).

Limbless Skink
Ophiomorus punctatissimus
Skinks SCINCIDAE

> 15–20 cm
> Skinks
> Back brown with fine longitudinal rows of spots;
  snake-like, small, without legs; 18–20 rows of scales

Description Small, snake-like skink with short and pointed head that might appear square in cross-section if seen from above, no legs. Tail about as long as body and quite broad with a rounded tip. Upperside glossy, yellowish brown or brown, with several longitudinal rows of small brown spots, each of which lies on one scale; lines on the tail more obvious than on the body. Underside pale and plain. About 18–20 rows of scales around mid-body.

Distribution and habitat Found in scattered locations in southern Greece (especially the Peloponnese) and Turkey up to an altitude of 600 m. On sunny, dry and warm, sparsely vegetated fallow land and clay loam soil covered by large stones.

Notes Lives mostly underground and is only very rarely seen. Most likely to be found in spring under stones. The biology of this species little known.

Similar species See Slow Worm, p. 196.

Slow Worm
Anguis fragilis
Slow worms ANGUIDAE

> 30–50 cm
> Slow worms
> Back brown, often with spots; snake-like, large, limbless; 23–30 rows of scales around mid-body

Description Smooth-scaled, snake-like anguid without limbs, with a blunt-ended tail that is marginally longer than the body and quite fragile. When damaged it hardly regenerates and is thus often smaller than the body. Head lizard-like, blunt-snouted and merges into body. Has closable eyelids (in contrast to snakes). Makes slow movements and may appear stiff. Upperside pale to dark brown, coppery, reddish or grey, mostly a bit brighter on the flanks. Adult females often have a dark central streak on the back and brown spots arranged in fine longitudinal lines or streaks. Males often have bright blue spots on the back (especially in eastern Europe, but not in Britain), which may become more intense during the mating season. Underside black to bluish grey, in males even yellowish.

Juveniles strikingly golden to glossy silver, with black central dorsal line, dark flanks and dark belly. 23–30 rows of scales around mid-body.

Distribution and habitat Found across Europe except for the extreme north, Ireland and southern Iberia. Also in north-west Africa, parts of south-west Asia. Occurs from lowlands up to alpine regions, in the Spanish Pyrenees and the Balkans up to 2,400 m. Widely distributed and common, especially in moist habitats with dense vegetation, e.g. meadows, forest edges, heaths and moors, also gardens, parks, fallow land. Under dead wood and stones.

Notes One of the most adaptable European reptiles. Active during the day from dawn; secretive and avoids strong sunlight. Diet: worms, slugs, spiders and insects. Hibernates from October–April in north of range (often 100 individuals hibernate together in deep holes in the ground), and mating takes place soon after. Female gives birth after three months to 8–20 young, which are about 8 cm long. Slow Worms can live up to 30 years.

Subspecies The nominate subspecies *A. f. fragilis* occurs in west and central Europe and *A. f. colchicus* in the east.

Similar species Limbless Skink is smaller (below 20 cm), with some longitudinal rows of brown spots, each of which lies

Female Slow Worm with newly hatched young

on a single scale, and with not more than 20 rows of scales around mid-body.

Additional species

■ **Peloponnese Slow Worm** *Anguis cephallonicus* (range in blue) is very dark and only inhabits the Peloponnese (Greece) and some neighbouring islands (Zakynthos, Cephalonia). It can be distinguished by the higher numbers of rows of scales around mid-body (34–36).

Peloponnese Slow Worm

European Glass Lizard
Pseudopus apodus
Slow worms ANGUIDAE

> 80–140 cm
> Glass lizards
> Very long, strong, snake-like; prominent groove on flanks; limbless; uniform brown

Description Very large and strong, limbless (except for tiny, barely visible vestiges of the hind legs) snake-like anguid with lizard-like, pointed head that merges smoothly into the body. Eyelids are closable. Due to bony plates, the skin appears armoured and has a groove that runs from the back of the head across the flanks to the base of the tail. The unbroken and pointed tail reaches 1–1.5 times the body length, but it may break and when it does so it barely regenerates. Upperside plain yellowish brown or chestnut with no pattern in adults. Sides of head and belly mostly brighter than the upperside. Juveniles grey with prominent, irregular, dark stripes, which disappear after 2–3 years. The keeled back scales become smoother with age (except for those on the tail).

Distribution and habitat In the coastal area of the Balkan Peninsula, also in Turkey and east to central Asia. Found from lowlands to higher altitudes, outside of Europe up to 2,400 m. Quite common, especially in sunny, dry or moist habitats with dense vegetation; often on stony slopes with plenty of hiding places among boulders, but also on cultivated land close to settlements, in vineyards or around stone walls.

Notes Classified in the genus *Ophisaurus* until recently. Diurnal species that likes to sunbathe, but may also be active during rain. When in danger flees noisily and may fall a few metres down steep slopes. Captured individuals do not usually bite, but they try to escape with strong winding movements. Diet: slugs and snails, worms and larger insects, such as beetles, but also young mammals and birds' eggs. Hibernates from October to March, after which mating takes place. The female deposits

A European Glass Lizard peers warily from a safe hiding place

6–12 soft-shelled eggs in June or July. Juveniles (10–12 cm long) hatch about 1.5 months later.

Subspecies The Balkan population belongs to the subspecies *P. a. thracius*, the nominate subspecies is from the Caucasus.

A relatively dark individual

Iberian Worm Lizard
Blanus cinereus
Worm lizards AMPHISBAENIDAE

> 20–28 cm
> Worm lizards
> Body small, snake-like; regular, ring-like grooves across body; limbless; uniform brown

Description Small, snake- or earthworm-like worm lizard without legs. The body has many grooves that separate rings of small squarish scales. Has a groove that runs along the flanks. Small head with tiny eyes separated from the rest of the body by a ring-like groove. Head can easily be mistaken for the tail. Upperside grey, brown, flesh-coloured or reddish brown, belly a little brighter.
Distribution and habitat Southern and central Iberian Peninsula,

found in lowlands and hilly countryside up to 1,800 m. On dry, loose, often sandy ground, but also on moist, humus soil and cultivated fields.
Notes Specialized species that lives underground. Occasionally found under stones. Diet: ants, ant larvae, other small insects. Lays 1–2 eggs in the summer.
Additional species
■ **Anatolian Worm Lizard** *Blanus strauchi* (range in blue) is similar. Found in Turkey and on Greek islands of Kos and Rhodes.

Worm Snake
Typhlops vermicularis
Worm snakes TYPHLOPIDAE

> 20–35 cm
> Worm snakes
> Body flesh-coloured, like a thin earthworm; scales on back and belly are the same size; very small black eyes

Description Small, specialized snake with thin body and regular, roof-tile-like scaling across the body. Appearance and movements like an earthworm. Small head not distinct from body. Has enlarged frontal scales on head and tiny black eyes. Short tail ending in a hard pin-shaped scale; this spiny tip is used for defence. Upperside leathery, shiny and flesh-coloured, belly brighter.

Distribution and habitat Found in the southern Balkans, at lower and middle altitudes. On open, dry, grassy fields with stones, under which the animals may be found especially in spring after rainfall.

Notes Has weakly developed eyes. Lives underground in tunnel systems dug by itself. As the mouth can barely be opened only small prey can be caught, e.g. termites, ants and their larvae, other small insects and centipedes. Clutch consists of 4–8 white eggs deposited in the ground in May.

Sand Boa
Eryx jaculus
Sand boas BOIDAE

> 40–80 cm
> Sand boas
> Pupils vertical, slit-like; scales on belly narrower than on rest of body; rudimentary hind legs; wide snout-scale

Description Stout snake with rudiments of hind legs and pelvis. Small head that is not distinct from the body with broad, triangular snout-scale (a tool for digging). Tail short, width of the ventral scales is only about one third of the body width. Tiny eyes with slit-like pupils. Upperside clay-coloured, grey or reddish brown with well-defined dark patches. The flanks have irregular dark patches. Belly grey to reddish with brown patches.

Distribution and habitat Southern Balkans in dry, open, stony lowland habitats, plains around rivers and fields close to open, sunlit forests. Lives under stones or underground in rodent burrows or tunnels dug by itself.
Notes Nocturnal. Kills prey such as mice, sleeping lizards and young birds by constriction. Viviparous, with 5–15 young born in late summer.
Subspecies *E. j. turcicus* in Europe, the nominate subspecies inhabits northern Africa.

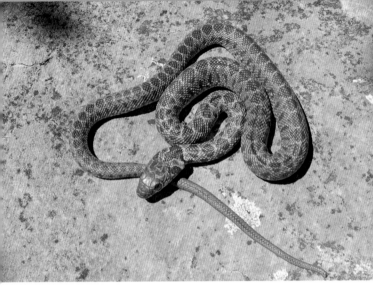

Horseshoe Whip Snake
Hemorrhois hippocrepis
Typical snakes COLUBRIDAE

> 140–200 cm
> Whip snakes (formerly genus *Coluber*)
> Horseshoe-shaped pattern on head; dark crescent between eyes; patches on back; 25–29 rows of scales

Description Stout body which appears slender due to long tail. Head distinct from body. Large eyes. Typical pattern on back of large, dark, often black-bordered patches on a yellowish to reddish background. Horseshoe-shaped pattern on head and dark bar between eyes. Small dark patches on flanks. Belly yellow to orange, some spots on the sides. Has 25–29 unkeeled dorsal scales around mid-body.

Distribution and habitat Iberia, Sardinia and Pantelleria, up to 1,750 m. Also Africa. Found in dry, hot, stony areas with bushes, e.g. scree and stony slopes.

Notes Agile, timid, diurnal. Lays 5–10 eggs in summer.

Similar species

■ **Coin-marked Snake**
Hemorrhois nummifer (range in blue) is a robust species which occurs in Turkey and on some Greek islands, e.g. Kos, Kalymnos.

■ **Algerian Whip Snake**
Hemorrhois algirus (range in green) was introduced from North Africa to Malta.

Dahl's Whip Snake
Platyceps najadum
Typical snakes COLUBRIDAE

> 80–135 cm
> Whip snakes (formerly genus *Coluber*)
> Body gracile; uniform brown with dark patches around the neck; 19 rows of scales

Description Very slender and gracile with a long tail. Head distinct from body. Large eyes and round pupils. Upperside uniform grey-green or olive to reddish-brown, becoming darker and browner towards the tail. Striking row of dark and brightly fringed patches around the sides of the neck. Belly plain whitish. Has 19 unkeeled dorsal scales around the centre of the back.

Distribution and habitat South Balkans to west Asia. Lowland and hilly country, in Armenia up to 2,500 m. Favours dry, stony places with some bushes, e.g. river plains, clearings, edges of trails and overgrown gardens.

Notes Very agile, diurnal. Diet: lizards, occasionally young mammals or crickets. Ground-dwelling but may climb well. Clutch of 3–5 eggs is deposited in May, young hatch in late summer.

Subspecies The nominate subspecies *P. n. najadum* inhabits the Caucasus, *P. n. dahli* is found in Europe.

Balkan Whip Snake
Hierophis gemonensis
Typical snakes COLUBRIDAE

> 80–110 cm
> Whip snakes (formerly genus *Coluber*)
> Dark patches on back spotted white; tail finely streaked;
> lower lateral scales spotted dark; 19 rows of scales

Description Slender with head distinct from body, large eyes and long tail that is often finely streaked. Upperside brown-, olive- or yellowish-grey with many small, dark patches interspersed with white streaks. Belly whitish, spotted dark on the sides. Has 19 unkeeled dorsal scales around mid- body.

Distribution and habitat Coastal area of the Balkans as far south as Greece, in lowland areas and hilly countryside. Found on sunny, stony, overgrown dry slopes, scree, or on stone walls and the borders of trails.

Notes Formerly classified as *H. laurenti*. Diurnal, agile, ground-dwelling snake that may climb well. Diet: lizards, mice, young birds, crickets. Clutch 4–10 eggs.

Similar species See Western Whip Snake, p. 206.

Additional species

■ **Gyaros Whip Snake** *Hierophis gyarosensis* (range in blue) is darker and closely resembles *H. viridiflavus*. It only occurs on the Greek island of Gyaros.

205

Western Whip Snake
Hierophis viridiflavus
Typical snakes COLUBRIDAE

> 120–180 cm
> Whip snakes (formerly genus *Coluber*)
> Reticulated yellow-green and black upperside; tail
  has fine longitudinal stripes; 19 rows of dorsal scales

Description Large, slender snake with distinct and elongated head, large eyes and round pupils. Long tail, which can be distinguished from rest of body by a pattern of fine, yellow-and-black longitudinal streaks. Rest of the body often has a strong yellow or yellowish reticulated or barred pattern on a black background. Head black, spotted yellow. Occasionally individuals with a completely black upperside can be found (formerly mistaken for separate subspecies, *Coluber v. carbonarius*). Juveniles olive to pale grey with a strong black pattern on head and dark patches, especially on the front of the body. Belly yellowish to grey, often spotted dark on the flanks. Usually has 19 unkeeled dorsal scales around mid-body.

Distribution and habitat Range from north-east Spain across France and Italy (including all large and many small Tyrhennian islands) to Croatia. Favours macchia and other scrub-like habitat, and dry slopes rich in vegetation. Also occurs in gardens, on forest edges, ruins and embankments, from lowlands up to 1,500 m. **Notes** An agile, diurnal and very fast snake. Diet: mainly lizards, young mammals and birds, but also other snakes, frogs, slugs and insects. Small prey is swallowed alive, whereas bigger prey is strangled before being swallowed. Mainly ground-dwelling but can climb well in scrub. Although harmless, when captured it will immediately try to bite in defence. Clutch consists of 5–15 eggs, which are deposited under stones or dead wood in the summer. Juveniles hatch after 1–2 months and feed mainly on small lizards and grasshoppers. Hibernates between October and March. **Similar species** Other whip snakes. Balkan Whip Snake is

An all-black individual: the 'carbonarius' variant

smaller and its front body shows many dark spots, which are streaked or spotted white in the centre; belly scales spotted dark. Caspian Whip Snake has uniform yellow or red belly scales, which are never spotted dark.

Western Whip Snake from Italy

Caspian Whip Snake
Dolichophis caspius
Typical snakes COLUBRIDAE

> 100–250 cm
> Whip snakes (formerly genus *Coluber*)
> Body and tail appear striped with fine longitudinal lines; belly plain; 19 rows of dorsal scales

Description Longest European snake, appears slender due to its elongated tail (maximum diameter about 4–5 cm). Individuals from populations on some Aegean islands are usually less than one metre in length. Relatively small head which is indistinct from the body. Large striking eyes with round pupils. Upperside of adult is mostly silvery grey to blue-grey or reddish to dark brown, with no striking patches. Individual scales are bordered black with a bright centre so that the complete body including the tail appears finely streaked. Head may appear reddish or orange. Juveniles have an obvious, bar-like pattern on the back, with the bars quite far apart and on a bright background. It does not show the spotted head pattern typical of Western Whip Snake, but is very similar in all other respects (see p. 206). Belly plain yellow, orange, brownish or black. Has 19 unkeeled dorsal scales around mid-body.

Distribution and habitat Southeast Balkans (not Peloponnese) and on many Aegean islands (not Crete or Rhodes) eastwards along the coast of the Black Sea to the Caucasus (not found in Turkey, as shown on map). An isolated population exists close to Budapest (Hungary). Found in lowland areas and hilly country up to 1,600 m, especially on sunny, stony and dry slopes with lots of scrub, or in open deciduous forests or areas with many thorny bushes.

Notes Diurnal, very agile and nimble hunter. Mainly ground-dwelling but may also climb quite well in bushes. Like all whip snakes very lively, if cornered it may jump up to half its body length and bite forcefully. It can be detected in the field by its often noisy escape if disturbed. Diet: mainly lizards and other snakes, also young birds, mammals and even large insects. After mating in

Head of Caspian Whip Snake

May 6–12 eggs are deposited in undisturbed places in the summer. The eggs hatch in September and the young feed mainly on grasshoppers and small reptiles, such as Snake-eyed Skinks. Hibernates from October to April.

Additional species

■ **Large Whip Snake** *D. jugularis* occurs in Turkey. It was formerly lumped with Caspian Whip Snake. In Europe only on the small Greek island of Agathonisi.

If cornered, Caspian Whip Snakes may jump at their attacker

Cat Snake
Telescopus fallax
Typical snakes COLUBRIDAE

> 60–100 cm
> Cat snakes
> Pupils vertical, slit-shaped; head clearly distinct from body; 17–21 rows of dorsal scales

Description Gracile, with flat, egg-shaped head distinct from body with trapezium-shaped cross-section. Eyes have vertical, slit-shaped pupils. Upperside pale grey, brownish or buff with dark patch on nape and bar-like patches on back and flanks, which can be inconspicuous. Belly whitish to yellowish with dark patches. Has 17–21 unkeeled dorsal scales around mid-body.

Distribution and habitat Adriatic coast from Istria to Greece, in stony lowlands and hilly country, especially on sunny and dry overgrown scree, dry stone walls and boulders.

Notes A primarily nocturnal snake. Diet: mainly sleeping lizards. Venom paralyses prey after 2–3 minutes. Grooved fangs at the back of the upper jaw (opistoglyphic). Humans hardly affected by the venom.

Subspecies Nominate subspecies inhabits most of the European range, *T. f. multisquamatus* and *T. f. pallidus* Crete and neighbouring islands.

Dwarf Snake
Eirenis modestus
Typical snakes COLUBRIDAE

> 50–60 cm
> Dwarf snakes
> Gracile; head not very distinct from body; brown, black collar and head markings; 17 rows of dorsal scales

Description Very thin and gracile snake with a small head that is not very distinct from the body. Has large round pupils. Upperside uniform pale brown to grey-brown, greenish or beige, typically with black collar and two black blotches on large head scales (often indistinct in adults). Belly glossy and whitish to yellowish. Has 17 dorsal scales around mid-body.

Distribution and habitat In Europe found only on a few Greek islands off the Turkish coast, for example Samos, Chios and Lesvos. In lowland and hilly country up to 1,800 m, especially on sunny, sparsely vegetated dry slopes or rocky scree.

Notes Diurnal and ground-dwelling. Leads a secretive life under stones. Diet: insects, scorpions and other arthropods, and also small lizards. Mating takes place in May and June, the small clutch consists of 3–5 longish eggs that are deposited under stones. The young snakes are just 8–10 cm long.

Smooth Snake
Coronella austriaca
Typical snakes COLUBRIDAE

> 50–90 cm
> Smooth snakes
> U-shaped patch on neck open towards back; stripe
  from nostril to neck; 19 rows of smooth dorsal scales

Description Quite small (rarely longer than 70 cm), but muscular and smooth-scaled snake with longish head not very distinct from body. Pupils round. Upperside variable, males usually indistinct pale grey or pale to reddish brown, females grey to black-brown. Head and neck has characteristic dark pattern consisting of a crown-like and more or less horseshoe-shaped marking which is open towards the back (*coronella* is the Latin for crown) as well as an obvious dark line running from the nostrils across the eyes towards the neck. There is also an indistinct line across the snout. Back has one or more longitudinal bands of bar-like spots which become darker towards the front of the body and may form a zigzag line (like an Adder's). Patches on the back are sometimes connected to form distinct bars or stripes. Some individuals have striking bright red markings bordered by dark lines or lack any pattern at all. Underside plain pale grey to almost black in females, reddish brown in males (sometimes finely spotted) and intense brick-red in juveniles. Head covered by large scales. Has 19 unkeeled dorsal scales around mid-body.

Distribution and habitat The most widely distributed of all European snakes apart from Adder and Grass Snake. Occurs from northern Portugal across wide parts of the southern and central Europe (including Scandinavia) towards the Caucasus and Urals. In Britain restricted to heathland in parts of southern England. Elsewhere, in the north of its range occurs at low and middle altitudes up to 1,800 m, while in Spain it favours hilly and mountainous country up to 2,700 m. Prefers sunny areas with lots of places to hide e.g. vineyards, dry grass-land, embankments, quarries or sunlit forests. Common in rocky habitats with dense vegetation and open places.

Notes Diurnal, quite slow and secretive. Active especially on warm, humid days or at dusk and dawn. Not poisonous or aggressive, but defends itself by biting. Diet: lizards, slow worms, small snakes and mammals. Juveniles feed mostly on lizards. Prey is strangled. Viviparous, mates in spring after hibernating from October to April; 3–14 young are born in late summer.

Subspecies Nominate subspecies

Smooth Snake in a typical rocky, dry grassland habitat

C. a. austriaca inhabits most of Europe, *C. a. fitzingeri* described from Sicily and south Italy and *C. a. acutirostris* from the Iberian Peninsula.

Similar species Southern Smooth Snake has the U-shaped head marking open towards front and a partially dark underside. False Smooth Snake has a vertical oval pupil and hood-like pattern on neck. Adder has keeled scales, vertical slit-like pupils and smaller scales on head (apart from a few bigger ones).

Individual showing typical stripe along the side of the head

Southern Smooth Snake
Coronella girondica
Typical snakes COLUBRIDAE

> 40–80 cm
> Smooth snakes
> U-shaped patch on neck open towards front; dark mark on snout; belly patchy; 21 rows of smooth dorsal scales

Description Very similar to Smooth Snake (see p. 212), but smaller and more slender (mostly only 50 cm long), with dark, horseshoe-shaped marking on head open towards the front. Striking dark line across the snout from eye to eye; lateral stripe on head reaches from the neck to the eye (not up to the nostril). Upperside yellowish, ochre, grey or reddish-brown with strong pattern of irregular dark patches and bars. Underside yellowish to orange-red.

Prominent dice-like pattern with black patches partially arranged in two rows. Has 21 unkeeled dorsal scales around mid-body.

Distribution and habitat Iberian Peninsula, southern France, Italy, Sardinia, Sicily and north-west Africa. Mostly at middle and lower altitudes, rarely up to 2,100 m. Sunny places such as scree, quarries, embankments and overgrown cultivated fields.

Notes Lays 5–10 eggs (in contrast Smooth Snake is viviparous).

False Smooth Snake
Macroprotodon cucullatus
Typical snakes COLUBRIDAE

> 40–65 cm
> False smooth snakes
> Vertical oval pupils; black collar-like marking on neck; belly spotted; 19–23 rows of smooth dorsal scales

Description Small, quite strong snake with flattened head not very distinct from body. Vertical pupils oval-shaped in bright light. Upperside brownish or pale grey with small, dark, often indistinct patches and a broad black collar around the nape. Underside yellowish to reddish with a black dice-like pattern. Has 19–23 unkeeled dorsal scales around mid-body.

Distribution and habitat South and south-west Iberian Peninsula, Balearic Islands and north Africa. Lowland and hilly country up to 1,500 m, especially in dry places with bushes, also close to humans in cultivated fields, dry-stone walls and ruins.

Notes Active night and dawn. Diet: geckos and sleeping lizards. Venom paralyses prey in a few seconds, but due to position of fangs in the back of the jaws there is little danger to humans.

Subspecies *M. c. ibericus* in Europe. The nominate subspecies and two additional subspecies can be found in Africa.

Four-lined Snake
Elaphe quatuorlineata
Typical snakes COLUBRIDAE

> 130–260 cm
> Typical snakes
> Back brown with four dark longitudinal lines; juveniles patchy; 25 rows of weakly keeled dorsal scales

Description Very large and strong snake, but rarely longer than 160 cm. Strong and elongated head with large scales clearly distinct from body. Pupils round. Upperside brownish, mostly ochre, yellowish-brown or grey to orange-brown with a strong pattern of four dark brown longitudinal stripes becoming indistinct towards rear. Dark lines along the sides of the plain brown head. Juveniles contrastingly patterned, without longitudinal stripes, but with obvious pale and dark markings on head. Body shows dark patches and barring. Underside yellowish, often spotted dark. Has 25 (rarely 23 or 27) weakly, but noticeably, keeled dorsal scales around mid-body.

Distribution and habitat Central and southern Italy (including Sicily), western parts of the Balkans along all of the Adriatic coast towards Greece. Warm, dry or moist lowlands and hilly countryside up to 1,400 m, especially in open deciduous woods and on rocky, well-vegetated slopes, embankments or dry-stone walls. Often close to wetlands.

Notes Rather lethargic, ground-dwelling snake that likes stony, scrubby habitat and is active during the day from dawn. Not poisonous; rarely bites when captured but may hiss threateningly. Diet: small mammals, birds or lizards, killed by strangling. The females lay 5–16 eggs under stones in July or August. In the central Italian village of Cucullo the Four-lined Snake is carried around in snake rituals to honour the holy founder of the Order of St Dominic.

Subspecies The nominate subspecies *E. q. quatuorlineata* inhabits almost all of the range. The smaller *E. q. muenteri* occurs on most of the Cycladic islands. *E. q. scyrensis*, the classification of which is still debated, occurs on Skyros. The almost stripeless *E. q. rechingeri* from Amorgos is sometimes regarded as a separate subspecies but possibly belongs to *E. q. scyrensis*.

Additional species

■ **Blotched Snake** *Elaphe sauromates* (distribution in blue) occurs from the eastern Balkans east to Turkmenistan and was regarded a subspecies of the Four-lined Snake until recently. Now it is recognised as a separate species. Instead of the longitudinal streaking it possesses a strong pattern of irregular, dark bars and oval patches which fade with age.

Immature Four-lined Snake showing transitional pattern of stripes and bars

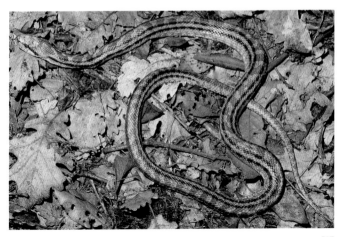

Aesculapian Snake
Zamenis longissimus
Typical snakes COLUBRIDAE

> 130–200 cm
> Typical snakes (formerly genus *Elaphe*)
> Body brownish to grey with fine whitish streaks;
>   23 rows of unkeeled dorsal scales

Description Large, strong, but slender and lithe snake that usually reaches 140 cm long, but can grow to 160 cm (males are often larger). Small, narrow head distinct from body. Round pupils. Upperside smooth and glossy, mostly fairly plain yellowish brown, olive or grey to black brown, sometimes with an indistinct line along the flanks. Many back scales and flank scales have a whitish outline which produces a finely streaked pattern. Juveniles have large dark patches on the back and two bright patches on the nape (similar to Grass Snake). Underside whitish or yellowish, but may be blue-black in very dark individuals. Has 23 (rarely 21) unkeeled dorsal scales around mid-body. Ventral scales on the sides are weakly keeled. **Distribution and habitat** Northern Spain across wide parts of central and southern Europe to Turkey. In Britain an introduced population exists in a small area of London. Found in both lowlands and hilly country, mostly below 1,000 m (rarely up to 2,000 m), often in areas with a warm and humid climate such as river valleys, open deciduous forests, forest edges and meadows on slopes with ivy scrub and bramble bushes, but also close to human settlements, for instance in gardens, among ruins or on embankments and overgrown stone walls.

Notes Mainly ground-dwelling, but also a skilled climber. Uses its ventral scales to help it to glide through bushes or along bark. Both diurnal and nocturnal. Has a strong bite but is not poisonous and not very aggressive. The snake depicted winding around the staff of Aesculapius, the symbol of the medical profession, is often considered to belong to this species. Diet: mainly small mammals such as mice and rats, as well as birds. Juveniles feed mainly on lizards. Larger prey is strangled before eating, smaller prey is suffocated by being eaten headfirst. Mating takes place in May and June after a hibernation period of about 4-5 months, with impressive wrestling fights between males that try press their opponent to the ground. In late July 5–12 eggs are deposited in decaying wood, humid soil or under stones.

Additional species

■ **Italian Aesculapian Snake**

Young Aesculapian Snakes possess pale markings on the neck which are similar to the Grass Snake's

Zamenis lineatus (distribution in blue) is grey on the underside. It was only recently split from the Aesculapian Snake as a separate species. It inhabits southern Italy (from the edge of the Aesculapian Snake's range), including Sicily north up to Naples and the Gargano Peninsula.

Adult Aesculapian Snake showing typical finely streaked pattern

Leopard Snake
Zamenis situla
Typical snakes COLUBRIDAE

> 70–100 cm
> Typical snakes (formerly genus *Elaphe*)
> Centre of back has large, reddish brown, black-fringed patches; 27 rows of unkeeled dorsal scales

Description Medium-sized, quite slender snake with conspicuous leopard-like pattern and small, elongated head which is not distinct from body. Pupils round. Basic coloration of upperparts yellowish, brownish, blue-grey or pale grey. Row of regularly arranged large orange-red to brown patches surrounded by black lines down centre of back. Smaller black spots on the flanks. Rarely the patches on the back are dumb-bell-shaped, divided in the middle, or stretched and merging in the middle. Head has bold pattern of two black, diagonal lines opening towards the rear. Underside yellowish at front of the body, darker on the central and rear part, becoming almost black. Has 27 (rarely 25) unkeeled dorsal scales around mid-body.

Distribution and habitat Southern Italy (including Sicily and Malta) and the western and south-eastern Balkans (including many Aegean islands as well as the Cyclades and Crete), east to Turkey. Found in lowlands up to 600 m (only rarely up to 1,000 m), in sunny habitats rich in vegetation and hiding places, which are dry, warm and moderately humid. Inhabits overgrown gardens or scree, open deciduous forests, ruins or dry stone walls, often close to streams, rivers or wells.

Notes Ground-dwelling, but also climbs well. Diurnal, slow moving snake that might also be active at dawn. Not poisonous and not very aggressive, but may bite strongly if captured. Shakes its tail tip when threatened. Diet: almost exclusively small mammals such as mice. The young snakes also feed on lizards. Mating occurs in April and May after a hibernation period lasting several months. In late August the young hatch from a small clutch of just 3–5 eggs. The hatchlings have an exceptional length of about 30 cm.

Similar species It is not likely to be confused due to its distinctive dorsal pattern. It is similar only to the juvenile Horseshoe Whip Snake, but the distributions of the two species do not overlap.

The Leopard Snake is one of the most colourful and distinctive of all European snakes

Ladder Snake
Rhinechis scalaris
Typical snakes COLUBRIDAE

> 100–160 cm
> Typical snakes (formerly genus *Elaphe*)
> Two dark longitudinal stripes on back; juveniles have
  ladder-like pattern; 27 rows of unkeeled dorsal scales

Description Large, quite strong snake (mostly smaller than 120 cm). Head not distinct from body and appears triangular due to the pointed snout. Pupils round. Upperside yellowish or dark- to reddish-brown with a well-developed pattern on the back consisting of two dark brown longitudinal stripes reaching down towards the tail. Juveniles have a weak line across the temples and a very characteristic ladder-like dorsal pattern (hence the name Ladder Snake) consisting of dark H-shaped patches, the crosswise lines of which fade with age. Underside whitish to yellowish with black spots in juveniles; in old animals usually plain silvery grey. Has 27 (occasionally 25 or 29) unkeeled dorsal scales around mid-body.

Distribution and habitat Iberian Peninsula (including Menorca) and southern France, from sea level up to 2,200 m. Especially favours warm and dry stony places rich in vegetation, for example overgrown vineyards, hedges, open oak forests, dry stone walls and even extensively used cultivated fields.

Notes Diurnal and crepuscular, particularly active at dusk and dawn. An agile and aggressive snake that bites energetically when captured but it is not poisonous. Ground-dwelling, but climbs well in bushes and on walls. Diet: mainly small mammals and birds; larger prey is strangled. Mating takes place in April and May after a hibernation period of 4–5 months. Females lay clutches of 5–25 eggs in the ground in July and August. The young hatch after 2–3 months. Hatchlings are between 10 and 25 cm long. The juveniles feed on lizards and larger insects such as grasshoppers and crickets.

Similar species Due to the characteristic pattern on the back of both the juvenile and

Almost full-grown Ladder Snake

the adult, this species is difficult to mistake, although juveniles can resemble the Southern Smooth Snake because of the barred dorsal pattern.

The characteristic 'ladder' pattern of the juvenile

Montpellier Snake
Malpolon monspessulanus
Typical snakes COLUBRIDAE

> 160–230 cm
> Montpellier snake
> 'Eagle eyed' due to raised eyebrows; elongated frontal scale, slender; 17–19 rows of unkeeled dorsal scales

Description Very large and strong snake appearing quite slender with narrow head that is not distinct from the body. Females generally not longer than 140 cm, males rarely longer than 190 cm. Protruding supraocular scales form a distinct rim and together with the large eyes they give an 'eagle eyed' appearance. Elongated frontal scale. Upperside more or less plain pale to dark grey, olive, black, grey or reddish brown. The flanks of the females often have dark spots which are edged whitish. Juveniles have a narrow, dark, brightly fringed pattern on the head and dark patches on the back. Underside whitish to yellowish, mostly spotted dark or marbled. Has 17–19 unkeeled dorsal scales around mid-body.

Distribution and habitat Found from northern Africa and Spain across southern France to the Balkan Peninsula (not on the Apennine Peninsula); also to Turkey and the Caucasus.

224

Found at lower and middle altitudes, rarely up to 2,200 m. An adaptable snake that prefers dry and warm habitats rich in vegetation: river plains, meadows on slopes with some bushes, forest edges, stone walls overgrown by scrub, embankments or cultivated fields.

Notes Diurnal, ground-dwelling, exceptionally agile and timid. Reacts early to danger by fleeing and will survey the area with an erect body posture. Often only the noise of the rapidly fleeing snake will be heard. Poisonous but the short opistoglyphic fangs are quite far back in the jaw. If threatened it will hiss energetically before biting. Venom very effective on lizards and snakes, small rodents and birds; causes minor swelling in humans. Mating occurs in April and May after hibernation. Females lay clutches of 1–18 eggs in humid soil in July and August.

Subspecies The nominate sub-

Montpellier Snake of the subspecies fuscus *from Greece*

species *M. m. monspessulanus* (which has 19 rows of dorsal scales) inhabits western Europe (up to north-west Italy), while *M. m. fuscus* (which has 17 rows of dorsal scales) inhabits the east of the European range (north-east Italy and the Balkans). A third subspecies *M. m. insignitus* occurs in parts of Africa and the Middle East.

Similar species Rarely mistaken because of its distinctive 'eagle eyed' appearance.

Half-grown individual of the nominate subspecies from Spain

Grass Snake
Natrix natrix
Typical snakes COLUBRIDAE

> 60–205 cm
> Water snakes
> Black and yellow broken collar on nape; nostrils pointed sideways; 19 rows of keeled dorsal scales

Description A rather large and strong snake with an oval head that is obviously distinct from the body. Round pupils and smooth head scales without any pattern. The length of the male is usually between 60–80 cm, females may grow to 80–140 cm; the largest individuals known (females) were about 180 cm long (in Switzerland) or 205 cm (on the Adriatic island of Krk). Diameter about 4–5 cm.

The upperside is generally grey, olive-grey, brownish-grey or black with a dorsal pattern that consists of black spots and patches in many subspecies, but in *N. n. persa* there are two bright longitudinal stripes and in *N. n. helvetica* there are broad, dark bars. The nape coloration in most subspecies (except for the Spanish *N. n. dastreptophora*, which is more or less uniform) shows a distinctive broken collar which is yellow, but sometimes white or orange-red towards the front and bordered black towards the rear. This may merge into a dark band along the nape. The underside is whitish grey with dark, square patches (like a chess board). Dorsal scales are keeled and usually number 19 around mid-body.

Distribution and habitat Found in almost all of Europe, in the north up to the 67th degree of latitude, but does not occur on some Mediterranean islands such as Crete, Malta or the Balearic Islands. Also in north-west Africa and in Asia east to Lake Baikal. Widely distributed in lowlands and mountainous regions up to 2,000 m in north and central Europe, and in the Spanish Sierra Nevada at heights up to 3,060 m. Found mostly in wet areas, in and near quiet or slow-flowing bodies of water

Grass Snake feigning death

such as lakes, ponds, streams and rivers. Also on moors, along rivers in forests, and in open deciduous forests, gardens, parks and quarries.

Notes Ground-dwelling snake, mainly diurnal but also active at dusk and dawn and only very rarely at night. Swims very well but is less aquatic than the closely related Viperine Snake and Dice Snake. Not poisonous,

Individual of the subspecies helvetica *from northern Italy*

*Grass Snake (subspecies
N. n. astreptophora) from Spain*

shy, quick to flee and does not usually bite, but releases a foul-smelling, yellowish liquid from its cloacal gland. Predated by raptors, herons, storks, crows, hedgehogs and certain fish such as perch, pike and catfish. When in danger it may feign death, lying on back with mouth open and tongue

*Grass Snake of the subspecies
sicula from Sicily*

hanging out, sometimes even showing blood. As soon as the threat is over the snake will escape. Diet: frogs, tadpoles, newts, fish, occasionally lizards and small mammals. In Britain mating takes place in April after emergence from hibernation in an insulated burrow. Sometimes large courtship gatherings numbering as many as 50 snakes may be found, in which males try to get the attention of the females. Clutches of 30–50 eggs are laid between July and August in decaying plants, piles of sawdust, compost or rotten tree trunks. In suitable places clutches from different females totalling 3,000 eggs have been found. **Subspecies** There are several subspecies of Grass Snake, some of which are still debated, and some hybrid forms between these subspecies can be found.

The nominate subspecies *N. n. natrix* is from north and central Europe (from Germany east of the Rhine to Russia), with *N. n. helvetica* from western Europe (Britain, France, Germany west of the Rhine, parts of Italy, Alpine countries), *N. n. astreptophora* from Spain and north Africa, *N. n. sicula* from south Italy and Sicily, *N. n. lanzai* from central Italy, *N. n. gotlandica* from Gotland (Sweden) and *N. n. persa* from the Balkans to the Caucasus. From the Cyclades islands, *N. n. schweizeri* inhabits Milos and *N. n. fusca* Kea. Two more island forms, *N. n. cetti* from Sardinia and *N. n. corsa* from Corsica, are considered by some authorities to relate to the the eastern (*N. n. natrix*) and a western (*N. n. helvetica*) subspecies respectively. **Similar species** The two other water snakes (subfamily Natricinae), the Dice Snake and

Black coloured individual with indistinct crescents on cheeks

the Viperine Snake, do not show the yellow-and-black pattern on the nape and have a different dorsal coloration. They are more dependent on water, their nostrils point upwards (Grass Snake's point sideways) and their scales are more strongly keeled (in Grass Snake the tail scales are often unkeeled).

*Corsican Grass Snake (*N. n. corsa*)*

Dice Snake
Natrix tesselata
Typical snakes COLUBRIDAE

> 60–150 cm
> Water snakes
> Nostrils point upwards; dark pattern on back;
> 19 rows of strongly keeled dorsal scales

Description Mid-sized, quite slender snake with elongated and slightly squarish head that is distinct from the body. Nostrils and eyes point upwards. The females are usually stronger and longer, and old females can reach 150 cm, but most individuals remain under 80 cm. Upperside brown, olive, yellowish, reddish, grey or almost black with a variable, dice-like pattern consisting of dark, mostly longitudinally arranged patches which some-times merge to form dark bars that alternate with brighter bars. Head indistinctly patterned, nape usually has a V-shaped mark. Underside whitish, yellowish or reddish with a dark chequered pattern or longitudinal stripes, sometimes even completely black. Dorsal scales strongly keeled, usually with 19 (rarely 17 or 21) rows around mid-body.
Distribution and habitat Southern and south-eastern Europe, from Italy north to the

Czech Republic and across the Balkans to central Asia. In the northern part of its distribution lives only in climatically suitable habitats in lowlands and hilly country, in the south occurs at altitudes of up to 2, 200 m. Prefers warmth and the vicinity of unspoiled larger bodies of water, especially rivers, lakes and quiet parts of streams with shallow areas and well-vegetated banks. Common on gravelly soils with bushes and dead wood. In some areas (e.g. the Bulgarian coast of the Black Sea) even in found in saline water, where it feeds on saltwater fish. **Notes** Diurnal, agile. Swims and dives extremely well. May stay under the surface for hours. Likes to sunbathe on bushes and stones that reach over the surface of the water. Not poisonous. A harmless snake that does not bite when caught, but

The Dice Snake is strongly associated with water and its diet consists mainly of fish. This individual is from Greece

may release a foul-smelling secretion from the cloacal gland. Diet: mainly fish, but also frogs, newts and tadpoles. The female deposits 5–25 eggs in humid ground or decaying plants after a long period of hibernation. The young hatch after 8–10 weeks. Hatchlings are 15–25 cm long.

Dice Snake from Lake Garda, Italy

Viperine Snake
Natrix maura
Typical snakes COLUBRIDAE

> 60–100 cm
> Water snakes
> Nostrils high up on the snout; zigzag pattern on back, ocelli on flanks; 21 rows of strongly keeled dorsal scales

Description Smallest of the three European water snakes. Has quite a strong body and a wide, slightly pointed head that is distinct from the body. Nostrils and eyes point upwards, pupils are round. Upperside of head covered by large scales. As with other snakes in the genus *Natrix*, is dependent on water. Ground colour brownish, yellowish, olive, orange-red or grey. Centre of back has a variable pattern of dark, viper-like zigzag markings, often broken into a two-row pattern of square patches. Flanks have a row of black patches, often in the form of ocelli with bright centres. In southern Spain also has two narrow, parallel running, yellowish longitudinal stripes along the back. On head has one or two dark, V-shaped marks that are open towards the back and a dark band along the temples. Underside whitish grey, yellowish or reddish, with a chequered pattern and square, dark brown patches.

Strongly keeled dorsal scales, usually in 21 (rarely 19 or 23) rows around mid-body.

Distribution and habitat Western Europe from Portugal across Spain and France to north-west Italy and south-west Switzerland. Also some Mediterranean islands (e.g. Sardinia, Balearics), north-west Africa and Atlantic islands. Found in a wide range of habitats: in or next to stagnant or slow-flowing bodies of water in lower and middle altitudes (in the Pyrenees up to 1,600 m), in open country (including cultivated areas), forests, banks of lakes, ponds, freshwater lagoons, streams and rivers.

Notes Diurnal, likes warm temperatures. One of the most common snakes on the Iberian Peninsula. Semi-aquatic, can swim and dive extremely well. The nostrils point upwards so can breathe while swimming. Diet: fish, amphibians and their larvae. When threatened responds by hissing, flattening the head and the body, pretending to strike with mouth closed and releasing foul-smelling

Viperine Snake with fish prey in the Pyrenees

secretion from the cloacal gland. Not poisonous despite its name, which only refers to impressive threatening behaviour and viper-like pattern on the back. Mating takes place early in spring after a hibernation period of 3–4 months. Between July and August, females lay 5–20 eggs into moist places in the ground or in decaying wood. After 40–45 days the young, which are 15–20 cm long, hatch and feed mainly on small fish.

Similar species See Grass Snake, p. 226. True vipers have vertical, slit-like pupils and small head scales.

Viperine Snake from Spain

Adder
Vipera berus
Vipers VIPERIDAE

> 50–85 cm
> Flat-snouted vipers
> Snout not upturned; pupils vertical; zigzag dorsal stripe; 21 rows of unkeeled dorsal scales

Description Strong viper but appears comparatively slender with short, stout tail. Females usually larger and stouter than males. Relatively narrow head hardly distinct from body and has a rounded (not upturned) snout. Pupils vertical and slit-like. Upperside very variable, ground-colour brown, coppery, blue-grey, black, yellowish, olive-green, orange or red; males usually more contrasting and greyer in overall appearance, females more reddish brown. Back has a dark zigzag line that usually lacks a black outline, in contrast to the similar Orsini's Viper (see p. 236). Flanks have dark spots or round patches. Has an X- or V-shaped marking on rear of head, the tip of which points towards the front. Also has a dark line on the temples that runs from the snout across the eye towards the neck. Belly grey, brownish or black, partly spotted white, underside of tail yellow or reddish. Upperside of head has small scales and one row of larger shields. Preanal scale undivided. Has 21 keeled dorsal scales around mid-body.

Distribution and habitat One of the most widely distributed snakes in the world. In Europe found from the Arctic Circle south to the Balkans, and in the Alps up to about 3,000 m. Range extends east to the Russian island of Sakhalin, north China and North Korea. Found in open and semi-open habitats, in humid places with a significant range of day and night temperatures, e.g. fringes of moors, heathland, scree, quarries, forest clearings or forest edges.

Notes Diurnal, timid and poisonous. Venom affects the blood and blood vessels. Bites are quite painful for humans but are very rarely lethal. Diet: mice, young birds, frogs and lizards. Hibernation in Britain is from mid-October to mid-March. Mating and fights between males (see Aesculapian Snake, p. 218) take place in April or May. Viviparous, gives birth to 5–20 young in late summer. In the Arctic Circle the annual period of activity is just 17 to 18 weeks.

Subspecies Classification of the subspecies is still debated. The nominate subspecies has the largest distribution (from Britain to East Asia). The

Adder is a very variable species and all-black individuals sometimes occur

smaller *V. b. bosniensis* of the Balkans is sometimes regarded a separate species. It tends to have a fading dorsal zigzag line.

Similar species See Asp Viper.

Additional species

■ **Seoane's Viper** *Vipera seoanei* (distribution in blue) replaces the Adder in Iberia and is rather similar to it and equally variable. The two subspecies, *V. s. seoanei* and *V. s. cantabrica*, inhabit the north of the Iberian Peninsula along the Cantabrian coast up to altitudes of 1,900 m.

Seoane's Viper from Iberia

Orsini's Viper
Vipera ursinii
Vipers VIPERIDAE

> 30–50 cm
> Flat-snouted vipers
> Snout not upturned; pupils vertical, slit-like; dorsal zigzag bordered black; 19 rows of keeled dorsal scales

Description Smallest European viper (usually less than 50 cm long); robust with narrow head that is hardly distinct from the body. Snout not upturned, pupils vertical and slit-like. Upperside similar to the larger Adder, mostly pale grey to yellowish-grey or brownish with a dark brown vertebral zigzag line clearly bordered black (in contrast to Adder, see p. 234). Has dark patches on flanks, a V-shaped head marking that points towards the front of the head and a dark line along the temples that reaches from the snout across the eyes towards the neck. Males are usually more contrasting. Belly grey, reddish or black, usually with dark patches, underside of the tail often yellowish or brownish. Has a row of large scales (and a few smaller ones) on the upperside of the head. Preanal scale unkeeled; has 19 strongly keeled dorsal scales around mid-body and thus appears quite rough.
Distribution and habitat One of

the rarest and most threatened European vipers; occurs only patchily in small, isolated populations in south-eastern France, central Italy, Hungary and parts of the Balkans. Extinct in Austria. Occurs especially at higher altitudes up to about 2,000 m (in the east also in lowlands), in sunny, open habitats like grassy slopes, high plateaux, scree and meadows with scattered bushes, often close to bodies of water or in moist valleys.

Notes Diurnal, but quite secretive and timid. Often uses rodent burrows. Venom less effective than other European viper species. Bites are quite painful for humans, but not lethal. Diet: large insects such as grasshoppers; also lizards and young mice. Mating between March and May after a hibernation period of about six months. Viviparous; the female produces 4–12 young in August or September.

Subspecies The classification of the subspecies is still debated, at present two European subspecies are accepted: the nominate subspecies *V. u. ursinii* from central Italy and *V. u. rakosiensis* from the lower Danube Valley from Hungary to Bulgaria. The acceptance of additional subspecies like *V. u. wettsteini* (south-eastern France), *V. u. macrops* (south-western and central Balkans) and *V. u. graeca* (north and

*Orsini's Viper (*V. u. ursinii*), Italy*

central Greece) is still debated.

Additional species

■ **Steppe Viper** *Vipera renardi* (range in blue, but occurrence in Moldova and Romania dubious) differs from Orsini's Viper in being larger and having 21 rows of dorsal scales around mid-body.

Orsini's Viper, south-east France

Asp Viper
Vipera aspis
Vipers VIPERIDAE

> 60–85 cm
> Nose-horned vipers
> Snout square and upturned; pupils vertical and slit-like; often barred; 21–23 rows of keeled dorsal scales

Description Mid-sized, stout viper with short tail and triangular head that is clearly distinct from body. Has upturned snout, but no horn on snout. Pupils vertical and slit-like. Upperside pale grey, yellowish, brownish or reddish to orange-brown, sometimes even black (especially in the Alps), with two rows of dark, staggered square patches or bars that sometimes merge into a zigzag stripe or a wavy line. Flanks have small or large dark patches that are not in line with the dark back pattern. Males are usually slightly longer and more contrastingly patterned than females. Underside grey, sometimes with dark patches, underside of tail yellowish or orange. upperside of head only covered by small scales. Preanal scale undivided; has 21–23 keeled dorsal scales around mid-body.

Distribution and habitat Parts of western and central Europe, from north-east Spain across France and Switzerland and throughout Italy. Found especially in hilly country at moderate altitudes, but in the Alps and Pyrenees up to 3,000 m. Favours sun-exposed, dry and vegetated mountain slopes, but also found in sunny river valleys, pastures, open forests or open areas interspersed with rocks and bushes.

Notes Diurnal and ground-dwelling with strong venom that is much more effective than that of the Adder (can be deadly for humans). Diet: mainly mice, also lizards and small birds. Mating takes place in April and May after a hibernation period of several months. The viviparous females give birth to 5–15 young in August or September.

Subspecies Five subspecies are currently recognized. These are: the nominate *V. a. aspis* from northern and central France; *V. a. zinnikeri* from the Spanish and French Pyrenees; *V. a. atra* from north-west Italy, south-east France and Switzerland;

Asp Viper of the subspecies hugyi *from Sicily*

V. a. hugyi from southern Italy and Sicily; and *V. a. francisciredi* from northern and central Italy. Apart from the last named, these were at one time all classified under *V. a. aspis*; the new taxonomy has not yet been fully accepted.

Similar species Adder and Orsini's Viper do not have an upturned snout, Lataste's Viper and Nose-horned Viper have an obvious horn on the snout.

Individual showing typical dark pattern on back and flanks

Nose-horned Viper
Vipera ammodytes
Vipers VIPERIDAE

> 60–100 cm
> Nose-horned vipers
> Scaly nose-horn; pupils vertical and slit-like; zigzag band; 21–23 rows of unkeeled dorsal scales, Balkans

Description Large, strong viper with a short tail and triangular head that is clearly distinct from the body. On the Aegean islands individuals rarely grow to more than 40–50 cm long. Snout has a striking, soft horn that is covered in scales. Pupils vertical and slit-like. Upperside white, pale to dark grey, yellowish, brown, reddish or black. Centre of the back has a relatively broad zigzag band, wavy line or rhombus-like pattern. Flanks often have dark patches or spots. Males are usually longer than females, and brighter with a more conspicuous pattern on the head and body. Belly grey with brown spots, underside of tail often yellowish, reddish or greenish. Upperside of head has many small scales and lacks large scales. Preanal scale undivided; has 21–23 keeled dorsal scales around mid-body.

Distribution and habitat Balkan Peninsula, from north-east Italy to Greece (including many Cycladic islands), as well as in

Turkey. Lowlands and mountain regions up to about 2,500 m, found on dry, sunny slopes and dry stone walls, in open deciduous forests, overgrown gardens or on scree with scattered bushes.

Notes Active during the day, at dusk and dawn. Mainly ground-dwelling, may climb between rocks and low bushes. A fairly quiet snake that usually flees from danger or relies on its camouflage. Only bites when absolutely necessary. Has one of the strongest venoms, which especially affects the blood and blood vessels and can be lethal to humans. Diet: mainly mice, also birds and lizards, other snakes and even large insects, like crickets and grasshoppers that are killed with a lethal bite. After a hibernation period that varies from 2–6 months depending on climate, mating takes place in spring and there are impressive fights between

Nose-horned Viper showing the distinctive snout-horn

the males (see Aesculapian Snake, p. 218). Viviparous; females produce 5–20 young between August and October.

Subspecies Six subspecies are known from the western part of the range, and the validity of some of these is not unanimously accepted. The subspecies are: *V. a. meridionalis* from the southern Balkans; *V. a. ruffoi* from the north Italian Alps; *V. a. ammodytes* and *V. a. montandoni* from the central and eastern Balkans; and *V. a. gregorwallneri* and *V. a. illyrica* from the north-western Balkans.

Nose-horned Viper from Greece

Lataste's Viper
Vipera latastei
Vipers VIPERIDAE

> 60–75 cm
> Nose-horned vipers
> Small snout-horn; pupils vertical; zigzag band along back; 21 rows of keeled dorsal scales; Iberia

Description Mid-sized, strong viper with a short tail and wide head that is obviously distinct from the body. Small snout-horn of 3–7 upturned scales. Pupils vertical and slit-like. Upperside grey, brown, occasionally reddish or black, with a dark (brownish in females), black-bordered zigzag or wavy central dorsal band. V-shaped head marking, flanks have dark patches. Underside grey with pale and dark patches. Small scales on upperside of head, large preocular scale. Preanal scale undivided. Has 21 keeled dorsal scales around mid-body.

Distribution and habitat Iberian Peninsula and north-west Africa, from sea-level to 1,800 m, rarely up to 3,000 m (Sierra Nevada, Spain). Sunny places rich in rocks and vegetation, and open deciduous forests.

Notes See Nose-horned Viper.

Subspecies Nominate subspecies *V. l. latasti* in north and east Iberia, *V. l. gaditana* in south-west Iberia and north-west Africa.

Ottoman Viper
Montivipera xanthina
Vipers VIPERIDAE

> 80–120 cm
> Oriental vipers
> Pupils vertical; two small and two large dark patches on head; wavy band; 23-25 rows of unkeeled dorsal scales

Description Large, strong viper with a broad, triangular head that is clearly distinct from body. Pupils vertical and slit-like. No nose-horn. Upperside pale grey, silvery white or brownish with dark brown, black-bordered rhombus or wavy central dorsal pattern. Typically has two small spots and two large, drop-shaped patches on head. Flanks barred dark. Belly pale grey, spotted dark. Head has small scales (except for preocular scales). Has 23-25 keeled dorsal scales around mid-body.

Distribution and habitat Turkey (up to 2,500 m) and some Greek islands such as Lesvos, Chios, Samos, Patmos and Kalymnos. In sunny, quite humid, rocky habitats rich in vegetation, open deciduous forests, cultivations.

Notes Very poisonous, can be fatal to humans.

Additional species
■ **Milos Viper** *Macrovipera schweizeri* (range in blue) is pale, not strongly patterned and very poisonous. Occurs only on Milos, Sifnos, Kimolos and Polynos.

243

Herpetological websites

AmphibiaWeb
http://amphibiaweb.org
Online access to information on amphibian declines, conservation, natural history, and taxonomy.

Amphibian and Reptile Groups of the UK
www.arg-uk.org.uk
A network of local groups involved with the conservation of amphibians and reptiles in the UK.

Amphibian Species of the World 5.3, an Online Reference
http://research.amnh.org/herpetology/amphibia/index.php
An up-to-date database hosted by the American Museum of Natural History.

Amphibians and Reptiles of Europe
www.herp.it
Photographs of European amphibians and reptiles, regularly updated.

Caudata
www.caudata.org
The longest running community for amphibian enthusiasts on the Internet, with a mission to promote learning and information exchange about newts and salamanders.

EuroTurtle
www.euroturtle.org
Sea turtle biology; distribution of the seven species of marine turtle; threats; conservation efforts; and identification keys.

Froglife
www.froglife.org
A British organization addressing key threats to amphibian and reptile survival including: destruction of habitats, disease, invasive species and persecution.

The Herpetological Conservation Trust
www.herpconstrust.org.uk
A UK-based charity and charitable company established to further the conservation of amphibians and reptiles.

International Reptile Conservation Foundation (IRCF)
www.ircf.org
A member-based organization that provides funding and practical support for reptile and ecosystem conservation.

Reptiles and Amphibians of France
www.herpfrance.com
A regularly updated site covering the herpetofauna of France. In English.

Save the Frogs
www.savethefrogs.com
An international team of scientists, educators, policymakers and naturalists dedicated to protecting amphibians.

Societas Europaea Herpetologica
www.seh-herpetology.org
A Pan-European herpetological network which aims to bring together herpetologists (neo- and palaeo-) and regional societies, in order to develop closer collaboration between them. Membership is open worldwide to individuals and institutions interested in the study of amphibians and reptiles, and their conservation.

The TIGR Reptile Database
www.reptile-database.org
A volunteer-run database providing primarily (scientific) names, synonyms, distributions and related data.

World Conservation Union (IUCN)
www.redlist.org/amphibians
A comprehensive assessment of the conservation status of the world's 6,000+ known species of frogs, toads, newts, salamanders, and caecilians.

Recordings of frog and toad calls

Recordings of frog and toad calls are available from the British Library National Sound Archive. They can be listened to via the Internet at www.bl.uk/listentonature/soundstax/frogs.html

Journals and magazines

A number of specialist journals and magazines cover reptiles and amphibians, among them:

Amphibia Reptilia (in English)
British Herpetological Society Bulletin (in English)
Die Eideschse (in German)
Herpetological Journal (in English)
Herpetozoa (in German)
Iguana (in English)
Nordisk Herpetologisk Forening (in Danish)
Revista Espana de Herpetologia (in Spanish)

Further reading

Arnold, E.N. & J.A. Burton (1983): *Pareys Reptilien- und Amphibien-führer Europas.* Hamburg, Berlin, Paul Parey (2nd edition). 270 pages.

Arnold, E.N. & D. Ovenden (2002): *A Field Guide to the Reptiles and Amphibians of Britain and Europe.* London, HarperCollins (2nd edition). 288 pages.

Böhme, W. (Hrsg.) (1981): *Handbuch der Reptilien und Amphibien Europas* - vol. 1: Echsen I. Wiesbaden, Akademische Verlagsgesellschaft. 520 pages.

Böhme, W. (Hrsg.) (1984): *Handbuch der Reptilien und Amphibien Europas* - vol. 2/I: Echsen II (Lacerta). Wiesbaden, Aula. 416 pages.

Böhme, W. (Hrsg.) (1986): *Handbuch der Reptilien und Amphibien Europas* - vol. 2/II: Echsen III (Podarcis). Wiesbaden, Aula. 436 pages.

Böhme, W. (Hrsg.) (1993): *Handbuch der Reptilien und Amphibien Europas* - vol. 3/I: Schlangen I. Wiesbaden, Aula. 480 pages.

Böhme, W. (Hrsg.) (1999): *Handbuch der Reptilien und Amphibien Europas* - vol. 3/II: Schlangen II. Wiesbaden, Aula. 350 pages.

Cabela, A., Grillitsch, H. & F. Tiedemann (2001): *Atlas zur Verbreitung und Ökologie der Amphibien und Reptilien in Österreich.* Wien, Umweltbundesamt. 880 pages.

Duguet, R. & F. Melki (Hrsg., ACEMAV coll.) (2003): *Les Amphibiens de France, Belgique et Luxembourg.* Mèze, France, Collection Parthénope, éditions Biotope. 480 pages.

Engelmann, W.-E., Fritzsche, J., Günther, R. & F. J. Obst (1993): *Lurche und Kriechtiere Europas.* Radebeul, Neumann (2nd edition). 440 pages.

Fritz, U. (Hrsg.) (2001): *Handbuch der Reptilien und Amphibien Europas.* Wiesbaden, Aula. vol. 3/IIIA: Schild-kröten I. Wiebelsheim, Aula. 596 pages.

Gasc, J.-P., Cabela, A., Crnobrnja-Isailovic, J., Dolmen, D., Grossenbacher, K., Haffner, P., Lescure, J., Martens, H., Martinez Rica, J.P., Maurin, H., Oliveira, M.E., Sofianidou, T.S., Veith, M. & A. Zuidervijk (Hrsg.) (1997): *Atlas of Amphibians and Reptiles in Europe.* Paris, Societas Europaea Herpetologica, Museum National d'Histoire Naturelle. 493 pages.

Greene, H.W. (1997) *Snakes: The Evolution of Mystery in Nature.* California University Press.

Grossenbacher, K. & B. Thiesmeier (Hrsg.) (1999): *Handbuch der Reptilien und Amphibien Europas.* vol. 4/I: Schwanzlurche I. Wiesbaden, Aula. 408 pages.

Grossenbacher, K. & B. Thiesmeier (Hrsg.) (2003): *Handbuch der Reptilien und Amphibien Europas.* vol. 4/II: Schwanzlurche IIA. Wiebelsheim, Aula. 350 pages.

Gruber, U. (1989): *Die Schlangen Europas und rund ums Mittelmeer.* Stuttgart, Franckh-Kosmos. 248 pages.

Günther, R. (Hrsg.) (1996): *Die Amphibien und Reptilien Deutschlands.* Jena, Stuttgart, Lübeck, Ulm, Gustav Fischer. 825 pages.

Hofer, U., Monney, J.-C. & G. Dusej (KARCH, Hrsg.) (2001): *Die Reptilien der Schweiz. Verbreitung, Lebensräume, Schutz.* Basel, Birkhäuser. 202 pages.

Kwet, A. & A. Schlüter (2002): *Frösche und Co: Froschlurche – Leben zwischen Land und Wasser.* Stuttgarter Beiträge zur Naturkunde, Serie C (Wissen für alle), Heft 51. 104 pages.

Malkmus, R. (2004): *Amphibians and Reptiles of Portugal, Madeira and the Azores-Archipelago.* Ruggell, A.R.G. Gantner. 448 pages.

Nöllert, A. & C. Nöllert (1992). *Die Amphibien Europas: Bestimmung - Gefährdung - Schutz.* Stuttgart, Franckh-Kosmos. 382 pages.

Pleguezuelos, J.M. Márquez, R. & M. Lizana (Hrsg.) (2002): *Atlas y Libro Rojo de los Anfibios y Reptiles de España.* Madrid, Asociación Herpetológica Española (2nd edition). 587 pages.

O'Shea, M. (2007): *Boas and Pythons of the World.* London, New Holland Publishers. 160 pages.

O'Shea, M. (2005): *Venomous Snakes of the World.* London, New Holland Publishers. 160 pages

Schreiber, E. (1912). *Herpetologia Europaea. Eine systematische Bearbeitung der Amphibien und Reptilien welche bisher in Europa aufgefunden sind.* Jena, Gustav Fischer Verlag (2. edition). 960 pages.

Thiesmeier, B. & K. Grossenbacher (Hrsg.) (2004): *Handbuch der Reptilien und Amphibien Europas.* vol.4/II: Schwanzlurche IIB. Wiebelsheim, Aula. 380 pages.

Zug, G.R., Vitt, L.J. and Caldwell, J.P. *Herpetology: An Introductory Biology of Amphibians and Reptiles.* (2nd edition). Academic Press.

Systematic list

Tailed amphibians

Salamanders and newts (Salamandridae)
<u>Land salamanders</u>
Fire Salamander *Salamandra salamandra*
Corsican Fire Salamander
Salamandra corsica
Luschan's Salamander
Lyciasalamandra helverseni
Alpine Salamander *Salamandra atra*
Lanza's Alpine Salamander
Salamandra lanzai
<u>Golden-striped salamander</u>
Golden-striped Salamander
Chioglossa lusitanica
<u>Spectacled salamander</u>
Spectacled Salamander
Salamandrina terdigitata
<u>Ribbed newts</u>
Sharp-ribbed Newt *Pleurodeles waltl*

<u>Brook newts</u>
Pyrenean Brook Newt *Euproctus asper*
Corsican Brook Newt *Euproctus montanus*
Sardinian Brook Newt
Euproctus platycephalus
<u>Pond newts</u>
Alpine Newt *Triturus alpestris*
Common Newt *Triturus vulgaris*
Palmate Newt *Triturus helveticus*
Montandon's Newt *Triturus montandoni*
Bosca's Newt *Triturus boscai*
Italian Newt *Triturus italicus*
Northern Crested Newt *Triturus cristatus*
Danube Crested Newt *Triturus dobrogicus*
Italian Crested Newt *Triturus carnifex*
Balkan Crested Newt *Triturus karelinii*
Marbled Newt *Triturus marmoratus*
Southern Marbled Newt *Triturus pygmaeus*
Lungless salamanders (Plethodontidae)
Italian Cave Salamander
Speleomantes italicus
Ambrosi's Cave Salamander

Speleomantes ambrosii
Strinati's Cave Salamander
  Speleomantes strinatii
Gené's Cave Salamander
  Speleomantes genei
Monte Albo Cave Salamander
  Speleomantes flavus
Supramontane Cave Salamander
  Speleomantes supramontis
Scented Cave Salamander
  Speleomantes imperialis
Olms (Proteidae)
Olm *Proteus anguinus*

Tailless amphibians

Fire-bellied toads (Bombinatoridae)
Fire-bellied Toad
  Bombina bombina
Yellow-bellied Toad
  Bombina variegata
Appenine Yellow-bellied Toad
  Bombina pachypus
**Midwife toads and painted frogs
(Discoglossidae)**
Midwife toads
Common Midwife Toad
  Alytes obstetricans
Iberian Midwife Toad
  Alytes cisternasii
Southern Midwife Toad
  Alytes dickhilleni
Mallorcan Midwife Toad
  Alytes muletensis
Painted frogs
West Iberian Painted Frog
  Discoglossus galganoi
East Iberian Painted Frog
  Discoglossus jeanneae
Painted Frog *Discoglossus pictus*
Corsican Painted Frog
  Discoglossus montalenti
Tyrrhenian Painted Frog
  Discoglossus sardus
Spadefoots (Pelobatidae)
Common Spadefoot
  Pelobates fuscus
Western Spadefoot
  Pelobates cultripes
Eastern Spadefoot
  Pelobates syriacus
Parsley Frogs (Pelodytidae)
Parsley Frog *Pelodytes punctatus*
Iberian Parsley Frog
  Pelodytes ibericus

True toads (Bufonidae)
Common Toad *Bufo bufo*
Natterjack *Bufo calamita*
Green Toad *Bufo viridis*
Tree frogs (Hylidae)
Common Tree Frog *Hyla arborea*
Stripeless Tree Frog *Hyla meridionalis*
Tyrrhenian Tree Frog *Hyla sarda*
Italian Tree Frog *Hyla intermedia*
Typical frogs (Ranidae)
Brown frogs
Common Frog *Rana temporaria*
Moor Frog *Rana arvalis*
Agile Frog *Rana dalmatina*
Italian Agile Frog *Rana latastei*
Italian Stream Frog *Rana italica*
Balkan Stream Frog *Rana graeca*
Iberian Frog *Rana iberica*
Pyrenean Frog *Rana pyrenaica*
Water frogs
Pool Frog *Rana lessonae*
Edible Frog *Rana* kl. *esculenta*
Marsh Frog *Rana ridibunda*
Greek Marsh Frog *Rana balcanica*
Epirus Water Frog *Rana epeirotica*
Albanian Pool Frog *Rana shqiperica*
Cretan Water Frog *Rana cretensis*
Karpathos Water Frog
  Rana cerigensis
Levant Water Frog *Rana bedriagae*
Iberian Water Frog *Rana perezi*
Graf's Hybrid Frog *Rana* kl. *grafi*
Italian Pool Frog *Rana bergeri*
Italian Hybrid Frog
  Rana kl. *hispanica*
American Bullfrog *Rana catesbeiana*

Tortoises and turtles

Tortoises (Testudinidae)
Hermann's Tortoise
  Testudo hermanni
Spur-thighed Tortoise
  Testudo graeca
Marginated Tortoise
  Testudo marginata
Weissinger's Tortoise
  Testudo weissingeri
Old World terrapins (Bataguridae)
Spanish Terrapin *Mauremys leprosa*
Balkan Terrapin *Mauremys rivulata*
Pond terrapins (Emydidae)
European Pond Terrapin
  Emys orbicularis

Red-eared Terrapin *Trachemys scripta*
Sea turtles (Cheloniidae)
Loggerhead Turtle
  Caretta caretta
Kemp's Ridley Turtle
  Lepidochelys kempii
Green Turtle
  Chelonia mydas
Leathery Turtle
  Dermochelys coriacea

Lizards

Agamas (Agamidae)
Starred Agama *Laudakia stellio*
Chameleons (Chamaeleonidae)
Mediterranean Chameleon
  Chamaeleo chamaeleon
African Chameleon
  Chamaeleo africanus
Geckos (Gekkonidae)
Moorish Gecko *Tarentola mauritanica*
Turkish Gecko *Hemidactylus turcicus*
Kotschy's Gecko *Cyrtopodion kotschyi*
European Leaf-toed Gecko
  Euleptes europaea
Typical lizards (Lacertidae)
<u>Green Lizards</u>
Sand Lizard *Lacerta agilis*
Eastern Green Lizard *Lacerta viridis*
Western Green Lizard
  Lacerta bilineata
Balkan Green Lizard
  Lacerta trilineata
Schreiber's Green Lizard
  Lacerta schreiberi
<u>Ocellated lizard</u>
Ocellated Lizard *Timon lepidus*
<u>Greek rock lizard</u>
Greek Rock Lizard
  Hellenolacerta graeca
<u>Typical rock lizards</u>
Bedriaga's Rock Lizard
  Archaeolacerta bedriagae
Mosor Rock Lizard
  Archaeolacerta mosorensis
Sharp-snouted Rock Lizard
  Archaeolacerta oxycephala
<u>Iberian rock lizards</u>
Horvath's Rock Lizard
  Iberolacerta horvathi
Aurelios Rock Lizard
  Iberolacerta aurelioi
Arán Rock Lizard
  Iberolacerta aranica

Pyrenean Rock Lizard
  Iberolacerta bonnali
Iberian Rock Lizard
  Iberolacerta monticola
León Rock Lizard
  Iberolacerta galani
Spanish Rock Lizard
  Iberolacerta cyreni
Peña de Francia Rock Lizard
  Iberolacerta martinezricai
<u>Wall lizards</u>
Common Wall Lizard
  Podarcis muralis
Iberian Wall Lizard
  Podarcis hispanica
Columbretes Wall Lizard
  Podarcis atrata
Bocage's Wall Lizard
  Podarcis bocagei
Carbonell's Wall Lizard
  Podarcis carbonelli
Ibiza Wall Lizard
  Podarcis pityusensis
Lilford's Wall Lizard
  Podarcis lilfordi
Moroccan Rock Lizard
  Teira perspicillata
Madeira Lizard *Teira dugesii*
Tyrrhenian Wall Lizard
  Podarcis tiliguerta
Sicilian Wall Lizard
  Podarcis wagleriana
Maltese Wall Lizard
  Podarcis filfolensis
Italian Wall Lizard
  Podarcis sicula
Aeolian Wall Lizard
  Podarcis raffonei
Dalmatian Wall Lizard
  Podarcis melisellensis
Balkan Wall Lizard
  Podarcis taurica
Skyros Wall Lizard *Podarcis gaigeae*
Erhard's Wall Lizard
  Podarcis erhardii
Milos Wall Lizard *Podarcis milensis*
Anatolian Wall Lizard
  Podarcis anatolica
Peloponnese Wall Lizard
  Podarcis peloponnesiaca
<u>Meadow lizard</u>
Meadow Lizard *Darevskia praticola*
<u>Common lizard</u>
Common Lizard *Zootoca vivipara*

Snake-eyed lizard
Snake-eyed Lizard *Ophisops elegans*
Psammodromus lizards
Large Psammodromus
  Psammodromus algirus
Spanish Psammodromus
  Psammodromus hispanicus
Spiny-footed lizard
Spiny-footed Lizard
  Acanthodactylus erythrurus
Algyroides lizards
Dalmatian Algyroides
  Algyroides nigropunctatus
Greek Algyroides
  Algyroides moreoticus
Pygmy Algyroides
  Algyroides fitzingeri
Spanish Algyroides
  Algyroides marchi
Skinks (Scincidae)
Bedriaga's Skink *Chalcides bedriagai*
Ocellated Skink *Chalcides ocellatus*
Levant Skink *Mabuya aurata*
Italian Three-toed Skink
  Chalcides chalcides
Western Three-toed Skink
  Chalcides striatus
Snake-eyed Skink
  Ablepharus kitaibelii
Limbless Skink
  Ophiomorus punctatissimus
Slow worms (Anguidae)
Slow Worm *Anguis fragilis*
Peloponnese Slow Worm
  Anguis cephallonicus
European Glass Lizard
  Pseudopus apodus
Worm lizards (Amphisbaenidae)
Iberian Worm Lizard *Blanus cinereus*
Anatolian Worm Lizard *Blanus strauchi*

Snakes

Worm snakes (Typhlopidae)
Worm Snake *Typhlops vermicularis*
Boas (Boidae)
Sand Boa *Eryx jaculus*
Typical snakes (Colubridae)
Whip snakes
Horseshoe Whip Snake
  Hemorrhois hippocrepis
Algerian Whip Snake
  Hemorrhois algirus
Coin-marked Snake
  Hemorrhois nummifer

Dahl's Whip Snake *Platyceps najadum*
Balkan Whip Snake
  Hierophis gemonensis
Gyaros Whip Snake
  Hierophis gyarosensis
Western Whip Snake
  Hierophis viridiflavus
Caspian Whip Snake
  Dolichophis caspius
Large Whip Snake *Dolichophis jugularis*
Cat snakes
Cat Snake *Telescopus fallax*
Dwarf snakes
Dwarf Snake *Eirenis modestus*
Smooth snakes
Smooth Snake *Coronella austriaca*
Southern Smooth Snake
  Coronella girondica
False Smooth Snakes
False Smooth Snake
  Macroprotodon cucullatus
Climbing snakes
Four-lined Snake
  Elaphe quatuorlineata
Blotched Snake
  Elaphe sauromates
Aesculapian Snake
  Zamenis longissimus
Italian Aesculapian Snake
  Zamenis lineatus
Leopard Snake *Zamenis situla*
Ladder Snake *Rhinechis scalaris*
Montpellier snake
Montpellier Snake
  Malpolon monspessulanus
Water snakes
Grass Snake *Natrix natrix*
Dice Snake *Natrix tesselata*
Viperine Snake *Natrix maura*
Vipers (Viperidae)
Flat-snouted vipers
Adder *Vipera berus*
Seoane's Viper *Vipera seoanei*
Orsini's Viper *Vipera ursinii*
Steppe Viper *Vipera renardi*
Nose-horned vipers
Asp Viper *Vipera aspis*
Nose-horned Viper *Vipera ammodytes*
Lataste's Viper *Vipera latasti*
Oriental vipers
Ottoman Viper
  Montivipera xanthina
Milos Viper
  Macrovipera schweizeri

Index